CONFRONTING THE CLIMATE

PALGRAVE STUDIES IN THE HISTORY OF
SCIENCE AND TECHNOLOGY

James Rodger Fleming (Colby College) and Roger D. Launius (National Air and Space Museum), Series Editors

This series presents original, high-quality, and accessible works at the cutting edge of scholarship within the history of science and technology. Books in the series aim to disseminate new knowledge and new perspectives about the history of science and technology, enhance and extend education, foster public understanding, and enrich cultural life. Collectively, these books will break down conventional lines of demarcation by incorporating historical perspectives into issues of current and ongoing concern, offering international and global perspectives on a variety of issues, and bridging the gap between historians and practicing scientists. In this way they advance scholarly conversation within and across traditional disciplines but also to help define new areas of intellectual endeavor.

Published by Palgrave Macmillan:

Continental Defense in the Eisenhower Era: Nuclear Antiaircraft Arms and the Cold War
By Christopher J. Bright

Confronting the Climate: British Airs and the Making of Environmental Medicine
By Vladimir Janković

Globalizing Polar Science: Reconsidering the International Polar and Geophysical Years
Edited by Roger D. Launius, James Rodger Fleming, and David H. DeVorkin

Confronting the Climate

British Airs and the Making of Environmental Medicine

Vladimir Janković

Contents

Illustrations

Acknowledgments

When I began working on this project, the idea was to explore the early history of environmental medicine and the ways in which thinking about human health meshed with the knowledge about the physical and social worlds in eighteenth-century Britain. Since then project went beyond either the history of medicine or that of earth sciences, and involved the ideas and practices that I now consider to be at the heart of modern environmental thinking, both in and outside medicine. In achieving this, I owe a great deal to colleagues, students, and friends who have helped extend my involvement in the history of environmental sciences, and who, through their own research, comments, and questions have encouraged and supported this work.

My first and foremost debt goes to the Wellcome Trust for the History of Medicine for funding this research through the University Award scheme at the Center for the History of Science, Technology and Medicine, University of Manchester. I wish to express my gratitude to all the staff and students in the center who during the years encouraged my work and helped me with research advice. Among those who came to assistance during my archival work, I would particularly like to thank the personnel of the Wellcome Trust Library in London, The British Library, The John Rylands Library, National Archives, and the Gallery of Costume at Platt Hall, Manchester.

Outside Manchester, I wish to express my gratitude to a host of individuals and institutions for giving me an opportunity to express my ideas and for receiving feedback. I am particularly grateful for scholars who took part in the Manchester meeting on 'City Health,' organized during the early stages of this project. In Britain, thanks also go to the conveners and participants of the seminars at the universities in Aberdeen, Cambridge, Leeds, Oxford, East Anglia, and Warwick. In Europe, I received valuable comments from biometeorologists at Garmisch-Partenkirchen and sociologists of medicine in Coimbra and Maastricht. In Brazil, I had opportunity to spend an extended teaching period at the Museu de Astronomia e Ciencias Afins. My gratitude goes to my host Christina Barboza and the Brazilian Ministry of Science. In the United States, I owe a great deal to the colleagues who met at the annual seminar at Center for Eighteenth Century Studies at Indiana University. My sincere thanks also go to the participants and commentators of my work at the Dibner History of Biology Seminar on Oceans and Atmospheres at Woods Hole. Finally, I am particularly indebted to the generous support of the Colby College Goldfarb Fellowship in Public Affairs and Civic Engagement, where I interacted with highly motivated students and had the opportunity to work with my host and wonderful colleague, Jim Fleming, who tirelessly rallied to my cause and gave excellent advice and support along the way.

Of the people who have been my closest readers and critics, I owe by far the greatest debt to Bill Luckin. Bill read several versions of the manuscript, and tremendously helped to clarify my ideas during the editing process. John Pickstone has been an inspiration all along, insightful as ever, and a great help in pointing to things I would otherwise miss. I am especially grateful to Kevis Goodman's friendship and unfaltering enthusiasm which gave me confidence and energy to complete. Many of my other colleagues commented on my presentations and early drafts, including Roger Cooter, Stephen Daniels, Georgina Endfield, Mary Favret, Tal Golan, Christopher Hamlin, Mark Harrison, Illana Lowy, Joan Mottram, Christopher Sellers, Tony Travis, Carsten Timmermann, Anne Villa, Dror Wahrman, and the anonymous reviewers. While I believe that all of those mentioned have contributed to an improved final manuscript, none is, of course, responsible for remaining weaknesses. Finally, I wish to thank the editorial team at Palgrave for making the publication process smooth and on time.

Of course, my deepest gratitude goes to my family, Kristina, Jovan, and Katarina, and my mother Olivera, for having to put up with years of research travel, revisions, and endless meditations on a favorite topic: the weather.

Introduction

One of the formative characteristics of modern life is the notion of environmental hazard. The notion derives from a separation between the realms we usually denote as the *inside* and the *outside*. In discussing environmental hazard, "inside" and "outside" can be seen as normative constructs rather than volumetric spaces: all *insides* share the fact that they are within their respective *outsides*, and all *outsides* share in being respective surroundings of their respective *insides*. But these relationships are neither tautological nor geometric. They rather reproduce a spatial imagery in which the inside has a *right* to be preserved from that which is outside. The historical understanding of disease in particular supported this dichotomy in considering bodily "insides" as something under the potentially hazardous influences of bodily "outsides." Classical humoralism endorsed this dialectic in framing disease as a result of exposure to incidental circumstances related to diet, locality, season, and lifestyle.

Today, concerns over the distribution of external health risks are natural as we strive to place ourselves in spaces conducive to safety, comfort, and health. We make decisions about where to live; how to dress; when to heat, cool, or light our homes; and how to avoid drafts and sunburn. In doing so, however, we continue to differentiate between "internal" health and "external" danger by seeking to balance the two. Yet while such concern may seem obvious to us, its origins and normativity reveal significant assumptions about how medical practice and theory define who we are in relation to our surroundings. What kept the distinction between the in and out so pervasive during a major part of medical and social history? Why was it (and still is) so obvious? And why did it take on such an importance during early modern times, when material improvements in everyday living would have assuaged the concern over external risk?[1]

One particularly important expression of this dichotomy was the eighteenth-century differentiation between the medical qualities of indoors and outdoors. A large portion of literature and practice concentrated on the conditions that, aside from their immediate causes, could also be brought into connection with places in which people lived. Early texts on occupational medicine are one example of this interest. Neo-Hippocratic research on spatial distribution and temporal changes of airs and waters is another example. But medical writers often became preoccupied with the diseases that, in their opinion, came about as a result of prolonged exposures to confined air and small-scale spaces such as the house. Confined air and indoor space became identified as causes of ailments whose progress or cure depended on a patient's ability to manage his or her lifestyle in the face of the aggressive urbanization and commercialization of leisure. Faced with unknown pathologies, medical practitioners turned toward lifestyles and urban spaces as explanatory frameworks. They deplored the sedentary life. They renounced debauchery. They advised that indulgence ruined robustness and thin clothing imperiled the lungs. Overheated parlors destroyed the body's natural immunity. In their scrutiny of everyday behavior, they negotiated a "claustrophobic" stereotype of indoors that provided (and still provides) a vision of health as endangered by unwise use of private space.[2] In short, as indoor life became more comfortable, it also became more detrimental to health.

In this book, I show that the lasting product of these and similar concerns and developments was modern "environmental medicine." It has often been noted that few issues stirred more debate in the eighteenth century than those of health and the physical surroundings.[3] Analyses of the connections between human ailments and physical surroundings spawned social action, leading to profound changes in the treatment of disease. There was a growing interest in Galenic non-naturals and Hippocratic "airs, waters and places." Public authorities laid out schemes to improve the hygienic conditions of public spaces. Investigation and action complemented each other, especially in crusades against stench and filth. Alain Corbin viewed this period as undergoing a major "perceptual revolution," during which both popular and medical attention turned toward understanding the pathological effects of circumstances previously outside systematic inquiry.[4] For example, research into the medical significance of marsh grounds, drying riverbeds, and decaying organic matter complemented investigations into bedchamber odors, kitchen vapors, wet linens, or faulty house insulation. Physicians studied urban air quality, medical topography, and altitude physiology. Naturalists kept weather records to trace the timing of epidemics,[5] described the pathological effects of tropical and polar colds, and analyzed the air in ships, prisons, and hospitals.⁶ Rumors about miasmas spawned health scares. Longer working hours stimulated the establishment of occupational medicine, while indoor entertainment changed the use and quality of residential space. Entrepreneurs responded by providing better sash windows, improved heating methods, and newly introduced ventilation. The general public gradually took to using umbrellas and rainproof garments. Some "changed the air" by traveling south.⁷

All such developments saw health and disease as a matter of how the body was placed vis-à-vis its surroundings.⁸ In eighteenth-century Britain, this topocentric approach became integral, not only to medical and architectural considerations, but also to enlightened humanitarianism, moral theory, and political argument. Living well and feeling healthy was identified with the properties of one's existential coordinates. The locus of illness moved from the individual body to the space between bodies. As a consequence, the healthy body came to be seen as an asset exposed to hazards, but that could also be trained to withstand them.⁹ "Inside" and "outside" defined the new fault lines in medical theory, health care, and early public health reformism.

The eighteenth-century medicalization of physical surroundings was not new, however. Similar attitudes had informed practices in the past devoted to the maintenance of everyday health, regardless of cultural circumstances or levels of social organization. Past architectural and planning practices, for example, testify to the care and expenditure invested in achieving a decent quality of life. Traditional estate surveying reveals the importance given to the construction process, the weathering of building materials, house orientation, altitude, and soil. Ancient Chinese urban layout was similarly based on an appreciation of physical setting, a tradition that survives in feng shui design methods. Early Japanese Tokugawa urbanism drew upon an understanding of wind paths; and *Airs, Waters, and Places,* the treatise attributed to Hippocrates, discussed street direction, sunshine, and prevailing winds. The Roman engineer Marcus Vitruvius proposed a network of city towers to gather information about wind directions as a basis of determining street alignment. During the Renaissance, climatic theory was embraced by city-builders such as Leonardo da Vinci, in his 1503 plan of Imola, while concerns about urban smoke and air quality are evident as early as medieval London, and in the 1661 publication of John Evelyn's *Fumifugium, or the Inconveniencies of the Aer and Smoak of London.*¹⁰

Yet, if such annoyances filled the space of daily life, past commentators saw them as unavoidable nuisances, to be shunned or ignored.

From the middle decades of the eighteenth century, however, they came to be conceptualized as health hazards. The definition of an "environmental hazard" was pivotal to all later thinking about health and physical surroundings. In that sense, what historians have identified as the "medical environmentalism" of the eighteenth century was interested not only in the objective predicates of experienced space, but also, in the words of Michael Dillon, in the norms of health that led to a "proliferating array of discourses of danger."[11]

Today, we differentiate between indoor and outdoor risks and seek to adjust our behavior to the conditions we expect in these realms. Yet, while such concerns may seem necessary to us, I show that their social and economic origins reveal significant assumptions about who we are, and, in particular, who we are in relation to our physical and social surroundings.

This book will explore the historical rationales behind the currently held view that "environment" represents one of the centerpieces of our health and identity. It will argue that the medicalization of lived space became possible only with the growth of affluence and material culture during the eighteenth century. It became possible only when the modern subject could resort to technological prosthetics to create a buffer zone between controlled comfort and random risk. Only at points in history when people achieve security, in a space sheltered from hazards such as leaky roofs, cold, dirt, and hunger, can they enjoy the privilege of reflecting on such misfortunes as avoidable threats rather than as intrinsic modalities of life. Even more importantly, only when it can afford to prevent unplanned suffering and emergencies can a human community blame such hardships for the existence of social, economic, and physical ills. Environmental threat, then, cannot be conceived before it is exteriorized into something that undermines peace, comfort, and health. In this sense, the beginnings of "environmental" attitudes lie in the separation between the experienced and the achievable, as related to a sense of general well-being. And it is also in this sense that eighteenth-century medical theory—with its prolific literature on the problem of "prolongation of life"—provides a crucial framework for thinking about the origins of modern environmentalism.

Medical interest in environmental hazard became widespread, particularly with the "invention of comfort." The technological ascent of comfort dislodged weather from its place in mundane experience. Before the "weatherproof" times, it is sometimes argued, bad weather was part of everyday communication and social movement. Improvements in housing, clothing, heating, and communication, however, led to "a detachment between Britons and the weather, which gathered pace throughout the nineteenth century...[i]n our time we have grown used to be able to do what we need or wish to do irrespective of what the weather is up to."[12] While this may be fair and accurate to a degree, does it explain the contradiction implicit in the soaring insurance claims in the wake of modern weather disasters? Does it reflect the ever-growing list of environmental hazards and chemical susceptibilities? Does it account for the apocalypse by climate change? Has not our technological flight from the weather only exposed our hubris and increased our frustration in the face of recalcitrant dangers?

For Jan Golinski, the agency that eighteenth-century naturalists assigned to nature in shaping social affairs did not eliminate the fact that civilization "remained vulnerable to atmospheric influences on human health that were often thought to be increasing in the conditions of modern society."[13] Despite the fact that during the eighteenth and nineteenth centuries, the weather and climate seemed to have become incidental rather than central to life, this development coincided with a growing medical concern over the ills attributed to airborne poisons, seasonal change, indoor pollution, and occupational and endemic exposures. It was precisely during this period that the medical profession became enthralled with the possibilities of climatotherapy, pneumatic chemistry, and other methods involving so-called "change of air" therapy.

These parallel yet seemingly contradictory trends suggest that the detachment from weather might in fact have instigated the medical study of external risks. Engineering a "weatherproof" lifestyle might have been the source of, rather than the solution to, health hazards, in that it raised expectations about environmental comfort. Indeed, comfort engendered discomfort. Dangers slipping through safety nets terrorize us in direct proportion to the strength of the failed nets. Hazards that elude prevention are perceived to be more threatening the greater our faith in the means to prevent them. In this book, I demonstrate that the perception of failed prevention acted as a key element in conceptualizing early "medical environmentalism" and, arguably, nineteenth-century public health reformism.

Eighteenth-century medical practitioners framed their investigations within their moral, aesthetic, and social value system. It was only from within this system *and* in relation to their expectations that they were able to define certain conditions as medically hazardous. Cultural norms provided the ground on which one could *assign* a medical meaning to nonmedical conditions (such as clothing and travel) or impute a risk to previously benign and pathologically neutral properties, such as drafts or candlelight. In other words, the construction of topocentric thinking about health was never simply a medical exercise. For theorists such as Mary Douglas and Aaron Wildawsky, discussion about the reality and magnitude of risk does not revolve around predictions about its "real" effects. Rather, it reflects the tendency of different groups to hold a vision of nature and society conceived in terms of their prior experiences of social cohesion and moral order. As there are no objective and value-neutral methods of measuring the reality of risk, what one believes about the magnitude of dangers is ultimately shaped by one's perspective of the future, which conforms to one's ideal way of life.[14]

True to this characterization, the majority of physicians discussed in this volume upheld the idea that environmental threat was contextual, not absolute. As humanitarians, who shared a belief in rational behavior and enlightened education, physicians understood risks as the physicochemical circumstances that *defied* medical foresight and *resisted* available methods of control and prevention. These anomalies, they also recognized, occurred in a society in which they *could*, in fact, be foreseen and prevented.[15] The remedy lay in education, improvement, and preparedness. In works on regimen and domestic hygiene, for example, illness due to the weather or foul air was framed as a personal failure to meet the norms of enlightened behavior. Living space was conducive to health only to the extent that it bore the marks of servant-based management, which made health one asset among others in the world of middle-class affluence. The middle-class house became a prime site of improvement, and the level of maintenance of its interiors conferred status on its residents. Elaborately maintained interiors set standards of ambient health and provided the baseline from which to negotiate risk as something alien to such space. This is why "neglect" became one of the key taboos in hygiene.

There was also a debate regarding new forms of eighteenth-century entertainment and institutional sites of social intercourse. As urban residents increasingly moved indoors to pursue experimental natural knowledge or enjoy intimate conviviality, their medical mentors railed against the ills caused by such trends. The key element of their critique was an assault on the aberrations associated with luxury, excess, and "artificial" modes of life exacerbated by the consumption of exotic foods. Atmospheric stress was claimed to present a particular threat to individuals indulging in a pampered, sheltered life. Such persons lost interest in outdoor pursuits, lowered their immunity through deskbound distraction, squandered their vitality in long night hours, and let fashion undermine their health. They preferred to look at the weather through a window. They traveled to the warmer climes of Europe. As medics began to get a clearer picture of these trends, it emerged that such routines led to the development of an unusual sensitivity to the physical circumstances of life. Those affected might possess subtlety for aesthetic expression, but they were also too easily "discomforted" by foul smells, coarse dress, strong foods, or damp hallways. Doctors reasoned that luxury led to indulgence, gluttony to lethargy, and an overall sluggishness to hypersensitivity. Again, the rise of comfort led to an epidemic spread of discomfort.

The moral assumptions about moderation provided a platform from which one could detect new, but culturally specific, dangers attributed to the combined action of the atmosphere and modern constitutional sensibility. This is not to say that the atmospheric risk was an invention solely of medical moralists and humanitarian pedagogues. The eighteenth century was a time of dramatic social change in England, during which towns and cities exercised a pull on the rural population and effected long-term demographic change. Urban growth was spectacular. It could not have failed to impress those observers who remembered quieter times. The degree of circulation of people and goods must have been beyond the ken of the average person. The physiognomy of towns changed—streets were laid out, buildings constructed, and parks created where previously none had been. This could not be ignored, from either a meteorological or medical point of view. Such physical manifestations of industry, transportation, and trade made a major difference to the quality of air and other properties in public spaces. The fact was asserted so universally that the period resembles our own in its anxiety over an "anthropogenic climate change."

For example, in one of the early descriptions of the "urban heat island," from 1761, an author asked why there had been such a high incidence of the common cold being contracted by country people immediately after their arrival in London. Warmed by "numberless buildings" and "multitude of fires," the author explained, the metropolitan air became rarefied and drew cooler country air to rush in, "as into a vacuum," creating "an artificial wind." The inflow was then scattered, deflected, and broken by the cityscape to circulate through lanes, allies, courts, and churchyards. Eddies assailed passersby in the street, but also penetrated houses, where the air crept in through walls and keyholes, underneath doors, and through ill-fitting windows.[16] Accounts like these portrayed atmospheric danger as a chaos of gradients, introducing to the public imagination a sort of medical "fragmentation" of space, whose risk levels depended on the demographic, architectural, and climatological features of social dynamics.

By "fragmentation of space" I mean the idea that the eighteenth-century medical "environment" was granular, segmented, hierarchical, and framed in terms of discomfort, anxiety, and neglect. It was a space "under construction," assembled in accordance with the medical readings of the relationship between hazard and discomfort. Medical authors presented this space as local and uneven in spread. They portrayed it as unpredictable and irregular in its effects. Some connected it with natural or stellar cycles, Hippocratic seasons, and miasmatic epidemiology. Others reduced it to the respiratory process and the chemical composition of gases. Soon a whole range of scales was in play: from the dangers associated with dress, to those in army barracks and hospitals, to the open spaces of streets, towns, and countryside.

In all such discussions, one particular issue claimed special attention: the medical meaning given to the distinction between "inside" and "outside." The dichotomy has been so universally adopted that it has been rendered invisible and has rarely been addressed in the history of medicine and related disciplines.[17] It was manifested in several idiomatic expressions. Most obviously, medical knowledge had traditionally distinguished between the internal operation of the body and what went on outside of it—the skin acting as a boundary. Second, the crossing of the boundary between indoors and outdoors became a medical concern in its own right. Thirdly, the distinction was made between those who lived inside their "native" climate and those who were forced to or chose to leave it, suffering as a result. And finally, the dichotomy could be applied to discriminate between what was closed and what was open, between what was structured and what was not: the city versus the country, orthodoxy versus heterodoxy, comfort versus discomfort, safety versus risk.

I assign this dichotomy high priority because it pervaded eighteenth-century medical literature. I also believe that before we attempt to explain the origins of "environmental medicine," it might make more sense to begin with explaining why the distinction between "in" and "out" served as the key formula in the representation of eighteenth-century causes of disease. This is crucial, because in all of its manifestations, the distinction between "in" and "out" was normative, rather than spatial. Inside one's skin partook of the quality of being "inside one's house," but also of being "inside one's neighborhood," inside one's village, or inside one's country.

In the normative sense, then, insides were represented as spaces of safety, stability, and control. They were surrounded by their respective outsides, represented as the domains of contingency and risk. Based on this dialectic, the ideas on how to feel safe, live well, or stay healthy derived in part from one's ability to manage the boundaries between the two domains and, more specifically, protect vulnerable insides from hazardous outsides. That the spatial understanding of eighteenth-century hazards derived from both the general fragmentation of space and the specific, normative division of space into inside and outside.

The logic of this tension is subject to more recent analysis of what constitutes risk more generally. While the notion of risk is subject to many readings, one argument is that risk is a function of hazard, exposure, and vulnerability. Risk can only be realized if its casualty is *exposed* to a hazard that can cause damage. In turn, damage can only come about if there is an inherent vulnerability in the exposed entity. Unless there is a relationship between all three components, there can be no risk. The changing patterns of risk can be seen to be associated with levels of vulnerability (lowered immunity), the nature of exposure (unhygienic dwellings), and properties of the environment itself. In eighteenth-century medical theory, as will be explained below, such correspondences were approximated by the functions assigned to the predisposing and exciting causes of disease.

This analysis asserts that a central constituent of risk is *exposure*. I use this term extensively in the rest of the book, partly because contemporaries used it, and partly because it brings us to the heart of their medical ideas and practice. As I will show, there are many contexts in which the term can be used; here I only want to bring to attention the fact that it provides the key entry into any discussion of environmental risk. Thinking in terms of the above dichotomy, it is commonly assumed that insides are always exposed to outsides. A safe interior is exposed to a hostile exterior. Cleanliness is exposed to dirt, bodies to cold, skin to the sun. Juxtapositions like these are so common in both medical and everyday parlance that it is surprising that they have so far eluded a systematic historical analysis. Part of the reason might be the fact that we have become accustomed to using the terminology of "environmental medicine" or "medical environmentalism." We have generally become used to referring to "the environment" as if it were nonproblematic, timeless, and self-explanatory. On closer inspection, and despite a scholarly interest in "environmental medicine," there has rarely been a discussion on what exactly is "environmental" about it. No contemporary authors refer to it as such. Neither physicians nor laypeople thought that they lived in *any* kind of "environment." And if there was no "environment," neither could there be an environmental medicine, understood as investigating "[t]he interactions between risk factors in the environment and human health."[18]

Today, "environment" is so ubiquitous that it rarely receives closer inspection. We live in an environment, whether biophysical, social, or financial. Yet if "environment"—as Einstein once quipped—is everything "that is not me," then it must refer to something that is at once anywhere and nowhere in particular. One may ask: the environment as opposed to what?[19] The difficulty is noted by geographers, who have suggested that the nonspecific nature of "environment" comes from its use as a category referring to "everything of relevance outside a specific demarcation."[20] And because environmental issues encompass "quite literally everything there is," there is a need to deconstruct the implicit assumption in order to understand the stakes behind the usage of the term.[21] Furthermore, for historians, taking environment for granted leads to presentism. In doing so, historians obfuscate the origins of the field, by assuming the seemingly objective existence of exterior risks in the sense in which it appears in modern scientific parlance. Steering clear of such tendencies, while keeping in mind the negotiated character of risk, I wish to avoid the temptation of speaking about the eighteenth-century medical profession as somehow introducing an "environmental paradigm" as a natural framework for dealing with large-scale phenomena such as epidemics and public health. Rather, following the French philosopher Georges Canguilhem, I prefer to emphasize the relational character of "milieu." Only to the extent to which one ignores that the milieu is something *"between two centers"*—such as two bodies or two organs or two locations—can the term lose its relational meaning and take on an absolute one.[22] In other words, if the environment and the body exist only relationally, the distinction between the two may be misleading, and any medical project based upon it pointless.

Caveats like these suggest that so-called "environmental medicine" needs a reconstruction before we can take for granted the presumed exchanges between the body and its surroundings. The approach proposed here is that we begin by replacing "environment" with "exposure." In so doing, I argue, we get closer to what the eighteenth-century physicians actually worried about when they talked about health in relation to airs, waters, and places. In this sense, the notion of exposure can be said to have both conceptually and historically predated the idea of environment. And, given its preponderance in medical and cultural context, it can be taken as the only proper, socially meaningful, framework of eighteenth-century thinking about health as dependent on the agencies *surrounding* the body. As I will argue throughout the book, central to this thinking about "exposure" was also thinking about the body's *vulnerability*, which, in turn, required protection by means of a neo-Galenic dietary regimen, domestic hygiene, and personal cleanliness. The constructions of risks and remedies were negotiated in parallel.

Perhaps the most well-known traditional expression of this vulnerability was to atmospheric change. The hackneyed view of English weather as prone to wild change can be traced to at least early modern documents,[23] but it was only during the eighteenth century that it became a medical, ethical, and economic problem. In medical literature, one finds a universal word of warning of the troubles arising from exposure to atmospheric change. At first reading, one is not sure whether the worry is about daily alterations, weekly shifts, or seasonal anomalies. It gradually becomes clear that the language is not precise: The reference is to any temporal scale, depending on the context. "In England we seldom enjoy any continuance of settled weather," wrote a medical pneumat and botanist from Guy's hospital, "as three fourths of our year may be regarded as no less disagreeable to the health, than disagreeable to our feelings."[24] "The variety of changes which happen in this island both in air and weather are so great and so sudden," another physician wrote, "that no precautions can possibly be taken to prevent the effect which it is obvious they must have on the state of perspiration." The succession of extreme heats and piercing colds undermined attempts to accommodate "our dress or provide suitable shelter to protect us in the various exigencies of the fluctuating seasons."[25] There being "nothing in Nature so changeable as the British climate," one worried how this is translated into a physiological condition. How did "rugged air," "variable climate," "unstable winds," "capricious seasons," "atmospheric alterations," and "changeable weather" influence a person's state of health and comfort? In other words, what was the medical fate of those "*exposed* to a sudden change of air"?[26]

It would be simplistic to see such statements as representative of some quintessentially "English" attitude toward the weather. Medical concerns over the effects of atmospheric change were ancient in origin and have been widely known to laypeople across Europe. Instead, I believe that the cliché confirms the crucial importance of the developments associated with the medical fragmentation of space, and the construction of risk, exposure, and vulnerability. It is therefore far more useful to interpret the presumed variableness of English weather as a vehicle for this negotiation. Such a quality embodied external risk, and provided physicians with a source of both pathological exposure and vulnerability. It was related to new ideas about health and society, rather than simply to the actual state of weather, which, as some believed, was not even that bad. "The weather in general is more temperate, and the transitions from cold to heat, and from heat to cold, take place in a much less degree in islands, then upon the continent."[27]

But the more specific reason for valorizing atmospheric *change* as "intrinsically" harmful lay, in my opinion, in the changing context of everyday life. Drawing on analyses of eighteenth-century regimens and domestic interiors, I argue that the bias toward "variability risk" conceals a culturally and economically driven interest in standards of comfort. The pathology assigned to ambiental *change* ought to be read as an opposition to the "healthiness" of ambiental *uniformity*—which, in this context, was the uniformity of indoor space.

The consequences were twofold. First, I argue that atmospheric risk to health was thought of in *opposition* to the comfort associated with the properties of urban middle-class households. Secondly, risks from a variable atmosphere became "natural" causes of disease only when physicians pronounced domestic space to be an "absolute" standard of comfort and health. As a result, "environmental" vulnerability was a ratification of the social status of such a space, whose thermodynamic properties, as is often pointed out, had to be reproduced by means of the conspicuous input (even waste) of labor. The contemporary conviction that the so-called "delicate constitution" faced a *greater level of risk* from changes in the weather was to say that they had got used to a *higher level of comfort*. And when late-eighteenth-century health reformers claimed that this infirmity arose as a consequence of luxury, they were confirming what the delicate classes wished to be acknowledged all along. Environmental disease ratified sensibility and social rank.

The book is divided into five chapters, each of which uses a contemporary medical issue to illustrate the exposure-oriented reasoning. As most of these issues drew from physiology, I begin by outlining the theoretical framework of research into the properties of air in relation to the human body. Here the emphasis will be on the French and Scottish schools of nervous sensibility, which I will interpret as legitimating the eighteenth-century cultures of "delicacy." One of the guiding concepts in the doctrines of these schools was "sympathy," which defined the mechanism of exchanges between the organism and its surroundings. But this mechanism also provided the rationale for presenting the modern body as extraordinarily vulnerable to the influences of its surroundings. I also use this preliminary chapter to argue that all assumptions about the state of health in relation to external risk derived from the axiomatic use of "exposure" as the key relational concept of early "environmental medicine."

Chapter 2 proceeds to link these issues with social trends outside the medical context. Specifically, I show that ambiental pathology owed its high profile in medical discourse to the criticism of a consumer society that placed a premium on physical comfort and fashionable behavior. Fashion, it was argued, encouraged individuals to pursue lifestyles that put them under a range of unhealthy exposures caused by nocturnal socializing, improper clothing, erratic diet, or polluted indoor air. The new forms of sociability and entertainment undermined the "natural" bodily constitution. The main targets of contemporary medical critique were the changes in the use, maintenance, and hygiene of domestic space. The increased need for the latter translated into a larger expenditure on the management of middle-class households. Domestic labor of this kind resulted in a production of standards of ambiental "normalcy" that made the dangers arising from an unhealthy "environment" equivalent to those arising from an "unmaintained" domestic space.

The arguments used to establish this correlation are analyzed in chapter 3. Here I explain that the discussions on what constituted aerial risk could not proceed without some knowledge of the means by which such risk could be prevented. Health hazards could emerge only to the extent that they could be contrasted with more benign alternatives. Accordingly, unhealthy interiors became so in virtue of the possibility of ventilated interiors. Ventilation thus became a medical issue, not simply because it solved the problem of foul air, but also because, at the same time, it helped physicians to construct foul air as a preventable cause of disease. Preventability engendered intolerability, not the other way around.

A similar dialectic underwrote the medicalization of clothing, I argue in the chapter 4. The debates over the health benefits of woolen garments next to the skin made sense only because such garments were already advertised as a solution to certain ailments. But the issue was also embroiled in the wider political economy. By the end of the eighteenth century, wool manufacturers fought against the expanding cotton industry by recruiting, among others, medical men to vouch for the thermal safety, even comfort, of flannel underwear. That the battle between wool and cotton unfolded on medical grounds was as important as the fact that, in the process, the medical profession came to define the importance of the smallest degree of atmospheric exposure and even to commercialize it. In medical tracts and trade journals, analyses of "intimate" climates expanded the domain of ambiental risk, but also made the rational choice of clothing a function of status and education.

The social and cultural implications of such arguments emerge fully in chapter 5, where I examine the scaled-up climatological risk associated with change of air and health travel. By the end of the eighteenth century, "the choice of air" became a major therapeutic option for those who could afford to believe that their condition would benefit from a travel to Spain or Italy. Individuals with serious conditions often traveled in hope of a last chance. In either case, they consulted physicians, who were often aware that not every condition benefited from a change of climate. To prevent blunder, they busied themselves with building a scientific taxonomy of resorts juxtaposed with a taxonomy of diseases. I examine how this knowledge built on existing meteorological views about the climate of the Mediterranean. But I also show its contentious character and, as in other chapters, interpret its evolution in the context of negotiated risk.

In identifying the physiology of air and atmospheric medicine as key sources of the medico-moral thinking during the late Enlightenment, this book investigates the origins of the distinction between the body and its meteorological surroundings on the eve of modern environmental thinking. But contrary to the common understanding of the "natural environment" as related to outdoor phenomena, this book also shows that weather as a medical entity remained only marginal with regard to the mainstream interest in the pathology of small-space conditions and local climates (such as in resorts). In this sense, I attempt to show that the Enlightenment understanding of exposures was not an expression of die-hard Hippocratism, but a result of the changes in life that brought about a vulnerability defined in respect to the bodily placements in space, time, and society. The aerial vulnerability naturalized the anxiety over exposures because both corresponded to the health as subject to external threat. In short, health became an insecure commodity under external risk, a commodity that could be maintained, augmented, or lost to the misrule of life.

Chapter 1

Exposed and Vulnerable

Ought the alarming Number of Suicides in this Country to be attributed to the Progress of Infidelity, Disappointment in the tender Passion, or any Peculiarity in our Soil and Climate?

Daily Advertiser, 1789[1]

By the early 1800s, the large volume of publications on the capacity of external factors to shape human physiology testified to a widespread consensus on the connection between an individual's state of health and the state of the air they breathed. Assumptions about the correlation between health and atmospheric events characterized the eighteenth-century neo-Hippocratic approach, which played a major role in the epidemiological analyses undertaken by medical practitioners and scholars such as James Jurin, John Huxham, and Thomas Short.[2] Charting the pathologies of seasonal change, their results set out the correlations between illness and weather.[3] These correlations indicated that certain diseases corresponded to atmospheric conditions, and that atmospheric conditions developed over the seasons. This link was therapeutically useful—it enabled the doctor to predict the timing, nature, and likely prognosis of the course of a disease. Physicians also knew that the spread of disease depended on fluctuations in the properties of soil, water, and air. They also argued that epidemics arose in places infested by malodorous mists—miasmas—which tended to occur during sultry weather and in the vicinity of decaying organic matter.[4] For many other ailments, physicians developed explanations involving bodily reactions to physical agencies such as heat, cold, wind, and humidity, regardless of the nature of atmospheric "constitutions" or the presence of miasmas. For example, damp situations were said to be *directly* responsible for colic, discharges, and consumption; cold air was directly responsible for rheumatism, sore throats, and watery eyes.

Eighteenth-century medical theorists framed these ideas in terms of the "external causes that affect Health," whose "irresistible impulse" had major implications for "animal economy" and "civil life."[5] Weather, climate, and health meshed in a variety of ways: from heat-induced headaches to endemic diseases, and from racial differences to forms of political rule. Virtually any physical, mental, or cultural trait *could* be related to aerial or terrestrial agency. People were thought of as inhabiting "native" climates in which the qualities of local air were as much about social identity as about medical conditions. When early modern geographers described local air as "congenial" or "insalubrious," they simply meant that such air supported or subverted health, not that it possessed qualities that, if isolated, could be found elsewhere: "[T]he surest mark of a good air, in any place, is the common longevity of its inhabitants."[6] Climate, health, and society were interlocked in a way that made any suffering part of the natural course of things: People were sick because they inhaled miasmas or were exposed to unhealthy vapors, smells, and winds. Bad air and rough weather were natural; sickness was, too.

This chapter shows how and why this belief atrophied during the eighteenth century. Physicians increasingly found that disease reflected an "unnatural" susceptibility to accidental changes in the physical space of everyday life. Although miasmas and seasons were still understood to undermine popular health, interest was growing in ailments associated with drastic *social*, rather than *natural*, change. "New" bad air infested social space, because bad taste, bad work, and bad habits shaped the lifestyle of new generations. The corruption of "natural" life eroded trust in a link between health and milieu. Air and weather conditions became unnatural; so, too, did sickness. Modern lifestyles produced more harm than good as civilization was turned into a perversion of cultural progress and providential harmony. This was the verdict of social and medical commentators, who demanded that any battle against the risks presented by "civilization" ought to entail ameliorative action at the sites of rapid social and material change. Health and disease were to be handled by curbing exposure to unnatural hazards in the social and physical surroundings.[7]

My aim in this and later chapters is to bring us closer to understanding the reasons behind this change. I argue that the mounting anxiety over new threats emerged from within contemporary physiological theory, but that its full expression came about as a result of the representation of physical space as one of permanent danger.

Thus, while traditional medicine talked about air as a disease carrier, the conviction that air might be the source of virtually *all* diseases emerged only in the mid-eighteenth century.[8] The origins and effects of this unprecedented expansion of the aerial threat lay at the heart of the emergence of environmental attitudes more generally. To understand how this was the case, it is not enough to demonstrate the prevalence of eighteenth-century research into the medical importance of seasons, constitutions, miasmas, and extreme cold as causes of disease. We need to go further, and ask how, in the first place, the nonpathological qualities of physical space became disease-bearing agencies. How did wind, rain, and sunlight enter into medical research and health regimens? Why the concern over what *might* become unhealthy, rather than with what already was? Why the fear of the *possibility* of disease, not disease per se? And finally, can we explain the emergence of an emphasis on the *atmosphere*—a term "that links the notion of 'mood' and 'ambience' to the air itself" and gives rise to the eighteenth-century "mental meteorology?"[9]

"Accidental Impressions"

Eighteenth-century physiology considered air to be intrinsic to all bodily processes. Drawing on sources going back to Hippocrates, Sanctorius, Thomas Sydenham, and Robert Boyle, physiologists regarded the air as an element that made up the body and enabled its operation.[10] Physiologists used studies in breathing, perspiration, and muscular action to establish a close correlation between the "animal economy" and air quality. Air supported life in a more profound way than other elements: "Men may live whole days without food; not a moment without air."[11] But physicians also recognized that air undermined life: "Distempers seldom arise from any other cause than the Air, for either it is too much or too little, or abounds with Infectious Filth."[12] The fact that life was sustained by mixing inorganic and living matter was one of the driving sources of contemporary medical practice. It saw the body as a fragile mix of matter and life, whose well-being depended on something as invisible and fickle the common air.

Air also contained water, dust, smells, miasmas, and marine salt. It contained oil, buttery particles, vegetable vapors, sweat droplets, volcanic ash, insect eggs, and maggots. Its weight on the ground was such that "it was surprizing, that every [change in atmospheric pressure] should not entirely break the frame of our Bodies to pieces."[13] According to calculations, every person lived under a pressure equivalent to the weight of thirty-two thousand pounds, which meant that a change in weather could amount to a two-ton change in pressure. Crossing the Simplon Pass on the way to the Italian Grand Tour could reduce the weight of air by almost four thousand kilograms. That such changes did not harm the human frame was due to the fact that the body and its atmospheric medium existed in an equilibrium—"such is the contrivance of infinite Wisdom"—that enabled them to communicate seamlessly in an exchange of fluids and gases. The body's permeability to air meant both survival and risk, and this recognition served as the framework for all subsequent thinking about meteorologically induced disease.

The fact that air enabled life, through processes that were as fundamental as they were unnoticeable, did not mean that its qualities were irrelevant to medical practice and social behavior. The fact that changes in the air were painless did not mean that they could not be felt. Neither did it mean that the sick endured such changes as well as the healthy. Failing to check the barometer and the weathercock could bring on a common cold to those caught in a shower or northeasterly wind. Refusing to adapt one's dress to the seasons lowered the body's immunity. Stale air in the bedroom caused lethargy and fatigue. Strenuous study by candlelight injured the eyes and lungs. Casual attitudes toward casual circumstances like these, it was said, perpetuated the error of treating health as a matter of random risk, common sense, and God's will. Physiological truths deserved better.

One of the reasons behind the emphasis on the medical meaning of everyday life derived from the early modern understanding of the body. By the early 1700s, the principles of medical practice were progressively seen as related to the experimental study of "animal economy." Following the work of William Harvey, Borelli, Robert Boyle, and Herman Boerhaave, early eighteenth-century medical philosophers conceptualized the body as a "hydraulic machine," subject to the quantifiable laws of matter in motion. This view precluded a dualist ontology, based on the separation of the body from its surroundings. The body was "thrown" into a space ruled by contingency, governed by laws no different from those governing the body's functions. The hydraulic body operated in a world in which it had to be maintained through vigilance, surveillance, and management. Being subject to the impact of heat, electricity, humidity, and gas elasticity, upkeep of the body-machine demanded an understanding of life-supporting processes subject to the laws of inorganic nature.

The possibility of risks to health arising from something as indispensable and erratic as common air had a striking consequence. It determined not only the level of doubt in providence, but also the degree of autonomy one had in a world made up of threats arising from continual change in the atmospheric surroundings. What was the meaning of somatic independence in a world of contingency known to shape organic processes? Anxiety over the possibility that health might depend totally on external contingency was one of the axioms of the mechanical conception of the body. In one of the earliest texts to explain the ways in which the air could *mechanically* injure the body, London practitioner Jeremiah Wainewright described the atmosphere as something that "constantly environs us, must needs impart its benign, or baneful *Influences* according to the various changes it undergoes."[14] Written in 1707, and in its fifth edition by 1737, his *Mechanical Account of the Non-naturals* outlined the iatromechanical reading of the body as a "machine" subject to the "Laws of Motion," whose influences on the animal economy could be understood in mathematical language.

Such a position reduced the medical art to the maintenance of a living machine by means of change of air, dieting, purging, sweating, cold baths, and the like. Such measures were more efficient than sporadic fixes by means of drugs and invasive treatment. Maintenance was especially appropriate for chronic disorders caused by low-level exposure to dust, humidity, cold, lack of sunlight, or impurities. Wainewright wrote that jaundice, asthma, and hypochondriasis were caused by changes in the density of "bodily" air, which, in individuals reportedly affected by changes in weather, caused a diminution in the cohesive force of the blood: "[S]ome people by their pain, can foretell any considerable change of the season, their blood being more rarefied against wet weather, or high winds, [that] will more forcibly press the sensible membranes whereby pains will be felt, they were free from before."[15] Such meteorological pathology was thus a result of involuntary response, not a prerogative of people endowed with a special "susceptibility" to the external world.

The association of air and the mechanical body was a dominant physiological assumption during the early eighteenth century. Part of the reason was the ascendancy of Newtonian physiology put forward by natural philosophers, such as Archibald Pitcairn, James Keill, Thomas Morgan, James Jurin, Henry Pemberton, and Stephen Hales. Their investigations adopted the mechanical framework, but insisted on the mathematical expression of their findings on musculatory action, motion of the lungs, blood circulation, and digestive pneumatics. Among those responsible for translating this view into medical practice were the Scots George Cheyne and John Arbuthnot. Adding relatively little to existing physiological theory, their importance rather lay in launching a comprehensive medical-ization of social space, through a combination of personal regimen and control of ambiental quality of life. Both men distrusted medical intervention: The use of drugs and traditional cure and surgery made sense from a professional perspective, but not from the perspective of a person interested in maintaining basic health. In their view, con-ventional medicine had little to contribute to such interests, since it intervened only when health had already broken down. Instead, the emphasis was to be on the qualities of one's surroundings, on which health *always* depended.

Cheyne, born in Aberdeenshire, became a leading expositor of Archibald Pitcairne's iatromathematics at Edinburgh University dur-ing the 1690s. Cheyne's *New Theory of Fevers* (1701) went through six editions before 1753. After becoming a fellow of the Royal Society, Cheyne developed practices in London and Bath, and went on to write on diet in *An Essay of Health and Long Life* (1724) and the *English Malady* (1733).[16] His own obesity and personal dietary trials led him to formulate the dietetic rules advertised in his many books. Seeing the body as "an Hydraulic Machin" made up of fluid-filled canals, Cheyne explained fevers as resulting from blockages in the flow of fluid.[17] In fevers associated with cold, air obstructed the work of the glands and stomach, and led to an abnormal flow of bodily liquids. Similar alterations were responsible for continual fevers in the tropics, which always followed a rapid change from sultry to cold weather. Workers on plantations in Maryland fell ill whenever they failed to dress according to the season, unaware that "the Stomach and Intestines being most Expos'd, and least Defended from the Cold Air, receive its first and strongest Impressions."[18] For Cheyne, the fact that in such circumstances the air *always* impinged on the body was the rationale for all further study in medical meteorology: "[T]he air is attracted and received into our Habit and mixed with our fluids every instant of our lives; so that any ill quality in the air so con-tinually introduced must in time produce fatal effects on the animal economy and therefore it will be of *utmost consequence* to everyone, to take care what kind of air it is they sleep and watch, breath and live in, and are perpetually receiving into the most intimate Union with the principles of Life."[19]

To redress the effects of such a comprehensively constructed risk-filled space, Cheyne developed a system of rules to help the weak, aged, and sedentary to pass "their lives in tolerable ease and quiet." He produced guidelines for each of the elements under examination—air, food, sleep, exercise, evacuation, and passions. The section on air looked into which of its properties had a role in such diverse circumstances as choosing one's house site, room layout in relation to winds, the use of drinks in cases of mild hypothermia, indoor heating standards, and general cleanliness. Cheyne warned the reader that London air had reached a dangerous level of pollution from coal fires, advising "pulmonicks" to get home before sunset and stay warm by the fireplace. Leaving no aspect of life untouched by his regimen, Cheyne nevertheless noted that his advice was not always heeded with the attention it deserved. Despite medical theory underlining the need for self-restraint, he noted, people were generally unlikely to act as if danger lurked in as many places as Cheyne's regimen implied. In real life, demand for restraint faltered before custom and pleasure, even when the consequences were more than obvious: "There is nothing more *ridiculous* than to see tender, *hysterical* and *vapourish* People, perpetually *complaining*, and yet perpetually *cramming*."[20]

Cheyne's friend, John Arbuthnot, shared the same worries. He concurred that the lack of appreciation of everyday risks was a worrying syndrome in contemporary society: "[S]urely the choice and measure of the material of which the whole body is composed, and what we take daily by pounds, is at least of as much importance, as of what we take seldom, and only by grains and spoonfuls." Arbuthnot—physician, satirist, and political commentator—was considered one of the more influential British men of letters in the early decades of the eighteenth century. Born near Stonehaven, on the northeast coast of Scotland, he was educated in mathematics and medicine, having graduated with an MD from the University of St. Andrews in 1696. During the 1710s, as a respected physician and fellow of the Royal Society, he also became a spiritual force of the "Scriblerus Club," whose members included Swift, Pope, Parnell, and Gay. Arbuthnot's first important medical study appeared in 1731, under the title *An Essay Concerning the Nature of Aliments.* Inspired by Cheyne's *Essay on Health and Long Life,* Arbuthnot planned it as the first of three volumes on health and longevity.

In switching the scope of his analysis from the exceptional to the common, and from the rare to the everyday, Arbuthnot followed Cheyne in making a connection between health and habit, and rejecting episodic therapy in favor of methodical self-care. The medical art lay in deterring mundane risks, rather than in curing their effects. This was true even if such a view might have been alien to "the reader [who] must not be surprised to find the most common and ordinary Facts taken notice of; in subject of this Nature there is no room for Invention." In such an art, the commonplace made all the difference, as "many important Consequences may be drawn from the Observation of the most common things." The "meanest things" demanded more attention than the " 'inventions' of medical art."[21] Arbuthnot confined himself to the task of taking the reader through the known branches of medical meteorology. He was content to raise public awareness of atmospheric risks, ranging from urban climates and terrestrial vapors to unseasonal change and pestilential constitutions. But he also applied this knowledge to the British context, providing advice on how to manage "nation-wide" risks. In so doing, he, like Cheyne, aspired to *domesticate* physiology as the key principle of self-care, and to use its results in the daily maintenance of private and public spaces.

In an appended list of aphorisms, he explained how to ward off natural threats, but he placed increasing emphasis on household activities as the mainspring of well-being: "[P]rivate houses ought to be perflated once a Day, by opening Doors and Windows, to blow off the Animal Steams." "Houses for the sake of Warmth fenc'd from Wind, and where the Carpenter's Work is so nice as to exclude all outward Air, are not the most wholesome." Protection against heat waves ought to include rest, shade, ventilation, "souterrains," and purpose-built grottoes. In acute disorders, "the Management of the Air in the Patient's Room" could be achieved by bringing in henbane, cowslips, and poppies to release curative vapors.[22] Going into minute detail, Arbuthnot raised the importance of mundane stimuli, expecting them to become the focus of medical prevention and social change—if for no other reason than the simple fact that he could think of "even more useless exercises."[23]

This understatement was not just an ironic twist of speech. In Arbuthnot's day, the nature of social behavior in everyday public space had been undergoing a major reordering. Following economic growth and the rising fortunes of the professional and middle ranks of society, the British were rapidly becoming aware of the magnitude, variety, and cultural importance of commodities in a "polite and commercial" society. The commercialization of books, news, games, fashions, clubs, and other sites of cultural exchange—followed by a massive consumption of everyday items, from food to dress to china—resulted in the development of lifestyles whose attendant risks became a target of social and moral criticism. In a society characterized by growing participation in public life and intercourse, *the everyday* was no longer a backdrop to social meaning usually associated with the display of princely power. If anything, display was now everywhere, communication frantic, and the pursuit of pleasure a norm. The proliferation of public life entailed a diversification of the somatic: People ran into medical trouble in proportion to their assiduousness in pursuing profession and pleasure. Sites of play and study also became sites of health and disease. The times became taxing, because the body ticked according to a natural, not a social, clock.[24]

The treatment of diseases that such bodies contracted required care, patience, and self-denial, which only a "few People are willing to undergo." It was a popular view that, while acute diseases cured themselves, chronic disorders remained uncured forever.[25] This might have been a logical view in a society in which medicine had traditionally meant intervention rather than prevention. But there were other reasons, too, why personal regime was not self-evident. It was time-consuming, expensive, and produced uncertain results. Its egotistic focus on individual discomforts might have been offensive in a world teeming with serious tribulations. Subsistence crises, the rising price of grain, epidemics, and snowy winters of scarcity seemed far more deserving of concern. It was insensitive to preach ventilation to the homeless. And yet, Wainewright, Cheyne, and Arbuthnot argued the opposite. In their view, worrying about small things was crucial. Medicine's objective, they wrote, was unrelated to the extraordinary and the episodic. Acute diseases were beyond control, epidemics spread despite precautions. Famines were deadly but rare. Most of the time, there was little that most people could do to prevent such calamities. In contrast to such events, prospects were not as bleak at the level of the everyday. Precaution at this level was crucial, action was necessary, and results were tangible. Medicine asserted itself as a system of precautionary "investments" aimed at achieving an optimum level of material quality of life, often encapsulated in the daily routines for the "prolongation of life."

Yet only those who believed that immediate restraint resulted in future benefit could take such investments seriously. Everyone knew that acting on a promise of future health lacked the incentive found in acting on the immediacy of pain. Before the advent of middle-class affluence and "the invention of comfort," to use John Crowley's expression, such actions were neither self-evident nor financially feasible. Wedding one's purse to discipline remained the prerogative only of those whose means kept them well above the level of survival and bad luck. In other words, concern over the atmospheric quality of life was class specific and signaled one's membership in at least an upper middle class. It thus served as one of the frameworks "for articulating the primary demands imposed by the conditions of existence upon men and women who sought seriously to preserve their physical well being."[26] And, as bodily management involved expenditure, environmental medicine had from its inception implied that health was a matter of "access to and control over the basic material and non-material resources that sustain and promote life at a high level of satisfaction."[27]

Vital Delicacy

The importance of the economic dimension in the acceptance of a medical regimen is underlined by developments that followed the initial discussions of Wainewright, Cheyne, and Arbuthnot. If the hydraulic body required maintenance by virtue of its mechanical makeup, the newly described "nervous" body required maintenance as it received, processed, and reacted to external stimuli. A medical regimen was, in this context, designed to help those fated to a lifestyle whose long-term effects produced a traumatic vulnerability to their physical surroundings, resulting in an eventual breakdown of healthy function. Cheyne was clear that his rules applied to the sedentary scholar, rather than the active farmer. Passive living and unproductive leisure enfeebled the body, increased delicacy, and brought on decrepitude. With the idea of delicacy in particular, social status had become inscribed in an "induced" weakness of the nerves.

A neurological reading of delicacy was implicit in Newtonian physiology. The medical historian Thomas Brown has shown that, by the 1720s, Boerhaave had been challenging the hydrological analogy popularized by Pitcairne's Edinburgh acolytes like Cheyne. By the next decade, however, the English Boerhaaveans, such as Stephen Hales and John Stuart, were emphasizing the role of solids rather than fluids in the functioning of the body. Their focus on fibers questioned the applicability of the mechanical approach, and highlighted the importance of muscular irritability in relation to external stimuli. This shift was already evident in attempts to explain the problem of so-called "elective purgation"—the fact that only living bodies showed a propensity to be electively purged by drugs. The idea rehabilitated the Galenic notion of *faculties*, which allowed animals to do things that inanimate objects could not: digestion, secretion, and production of blood. Furthermore, the mechanical account of animal heat was contradicted by experiments—friction, for example, could not resuscitate life. Gradually, medical interest moved toward the vital properties of organisms, which, in Brown's terminology, led a "subtle transformation in English physiology" toward fibers, convulsive motions, and—importantly—the stimulus response.

This transformation was evident in the language of medical men such as Bernard Lynch, a London practitioner. It was obvious to him that if the air's properties controlled the behavior of bodily solids, then the doctor's duty was to consider the state of the atmosphere as a key condition in diagnosis and etiology. For Lynch, the risk of living in colder regions was reflected in the fact that the local climate kept the body's fibers in "continual oscillatory motion," which caused "the tension of the fibres and the whole nervous system." In such geographic zones, the temper of the body differed from that found farther south, so that northern airs constricted and relaxed muscles: "[W]hen a warm weather suddenly succeeds the cold weather, it relaxes the fibres and the vessels destined to carry serum and lymph may admit the blood, which is an inflammatory state."[28] Yet such processes did not prevent other hazards, especially those arising from direct atmospheric influences: Cold air, when in contact with the surface of the lungs, stopped circulation and caused coughs; headaches arose after exposure to odorous particles; consumption was triggered by the lungs' contact with saline particles in local air. Health, in other words, was a state of controlled stimulation of fibers, whose state depended on exterior stimuli.

During the 1730s, interest in the phenomenon of stimulation signaled the rise of vitalist neurology, with significant implications for the conceptualization of the body vis-à-vis its atmospheric milieu.[29] Adapting Albrecht von Haller's theory of irritability to explain the assimilation of external influences by an innate "sensibility," medical theorists affirmed that life itself was a kind of susceptibility to external stimuli. This idea reinforced the "environmentalist" approach of preventive medicine, as it tied the inner organization of the body to weather and climate.[30] At the University of Medicine at Montpellier, Dr. Louis La Caze spoke of "the general external organ," whose function he defined as transmitting outside influences to the brain and the epigastric system. The organ involved the skin, the integuments, and the tissue under the skin, which acted as "a gate wide open to all external impressions."[31] From the vitalist viewpoint, the body was alive *because* it registered such impressions and reacted to them.[32] Furthermore, as Elizabeth Williams has recently explained, this approach allowed physicians:

> to claim as part of their purview *any substance*, general environmental condition, or activity that contributed to the maintenance of the life of the organism (all *ingesta*), formed its milieu (air, water, climate, the Hippocratic "places"), or conditioned its own response (exercise,

work, habits). Within such a rubric, *virtually anything* could be said to impinge on health or sickness and therefore to fall under the authority of the doctor.[33]

During the 1750s, as Laurence Brockliss and Colin Jones have argued, an "obsessive concern [with] air quality" reflected the belief that the "air was the source of virtually *all* diseases."[34] Such views replaced the idea of disease as a condition of the individual organism, with the idea of disease as a nervous reaction of the body exposed to a particular physical milieu.[35] By the mid-eighteenth century, the nervous system had become a key element of teaching at the medical professoriate of the University of Edinburgh.[36] Drawing on Lockean psychology and vitalism, the salient feature of the Scottish research into neural function[37] was its environmentalist bent, which required the existence of a reactive organism, endowed with susceptibility to the exterior.[38] The nervous system mediated between the organism and its surroundings, and nervous sensibility regulated the diet, exercise, climate, sleep, and passions. The main proponents of the doctrine were Robert Whytt, William Cullen, Alexander Monro *secundus*, and John and James Gregory.[39] During the 1730s, for example, Whytt criticized the Boerhaavean concept of the "inanimate machine" as incomplete and reductionist, and looked for ways to reintroduce the vital principle into physiological theory. He spoke of "the sentient principle," which let the body "feel" outside stimuli and allowed the body's functions to resonate with the medium: "[T]he living body is trembling, quivering, restless and moving about in the deepest and the smallest part of its fibres."[40] This made the nervous system the seat of sentient and intellectual action, and directed medical attention to factors such as occupation, climate, and topography: "Sensibility united environment, organism, and mind."[41]

Weather Marionettes

But sensibility had other meanings, too. Like the humoral theory of temperament, it was the "basis of a physiological anthropology of race or social class," as in the work of John Gregory and William Cullen.[42] It stood for discrimination and refinement. The savage's insensibility to bad weather and coarse food produced an insensibility to the plight of others: "[I]nhabitants of a rude and uncultivated climate [were insensible when compared to] those of a polished and civilised nation."[43] Sensibility was an emblem of rank, and a badge of superiority and accomplishment. Innate variations in its degrees were reflected in differences in social status. There were as many degrees of sensibility, wrote Cheyne, as there were "degrees of Intelligence and Perception in human Creatures." "Quick Thinkers" reacted to the slightest touch of a pin; the less sensible withstood being "run thro the Body" with a sword.[44] The "easiness of being acted upon by external Objects" betokened status and[45] conferred on the individual a delicacy with which to discern the finest changes in his surroundings.[46]

"Delicacy" became a key term in thinking about the vagaries of health. "Constitutional delicacy" and "delicate constitution" referred to inherited as well as acquired states. John Gregory, professor at the Institute of Medicine in Edinburgh, wrote in 1774 that women had a "natural softness and sensibility," which made them "susceptible

Figure 1 Richard Earlom (after George Romney), *Sensibility* (1789). Wellcome Trust Image Collection.

to the finer feelings."[47] But finer feelings, better education, and mental sharpness resulted in a "correspondent delicacy of bodily constitution," which made female receptivity comparable to that of the old, infirm, and wounded: "[A]lthough there are some individuals capable of bearing great changes of the air, there are others in whom the slightest alteration with respect to density has considerable effect."[48] Arthritis was a menacing barometer; vertigo announced rain.[49] Fashionable women suffering from chronic dizziness were virtually barred from outdoor activities for fear of collapse; headaches, languor, and quick pulse added to the plight of those unable "to raise [the head] from the pillow without suffering greatly."[50]

Delicacy referred to the fragile female body, but it also had a transgender quality. It indicated nicety, softness, indulgence, and, importantly, politeness and civility.[51] It was a "weakness of constitution" found in persons "apt to be ruffled, and put out of Humour by every little Accident, easily dejected by Disappointments."[52] "Being easily dejected" conformed to the vitalist idea of life as a recorder of everyday surroundings, which, from around the mid-eighteenth century, began to acquire a highly prominent, if increasingly negative, medical meaning. If physiology allowed trivial circumstances to dictate health and disease, danger loomed not only in the patient next door or the stench in the street, but also in the air itself. Yet, in addition to worrying about how the external atmosphere compressed the air trapped in the body's canals, the new doctors spoke of attrition of the nerves under exposure to atmospheric stimuli.[53] The nervous body absorbed the impact of weather through adjustments, which, in some cases, led to conditions as serious as asthma, arthritis, bronchitis, and consumption. More often, however, the result was the decline of a person's ability to fight weather-induced stress, "a particular weakness, delicacy and sensibility of the nerves," caused by sudden changes of air quality, errors in diet, fatigue, passions, or hard mental work.[54] Whytt referred readers to the case of a 28-year-old man who, after contracting a stomach disorder and mental depression, became "so *sensible* of any change of weather, that, by a general feeling of weakness and inactivity, and of pains in his joints, he could have told, in the morning before he got out of bed, that the weather was moist and rainy, or the winds easterly or southerly."[55]

References to extreme vulnerability to weather became a staple of medical histories. Delicate bodies were known as "human barometers," which literally registered the weather[56] via the sympathetic response of the nervous system. Physicians wrote of "alterations in the weather, by which the health of people is so remarkably affected";[57] of patients "agitated" prior to the arrival of a storm;[58] and of constitutions "peculiarly sensible to severities of the weather."[59] Practitioners were aware of the fact that there were "very few instances" of people who "pass through all the changes and vicissitudes of air and weather, without feeling any sensible disorder or inconvenience."[60] Statistics showed that as many as five out of six suicides took place at both ends of the winter, when "a gloomy atmosphere" "greatly affects the spirits."[61] But sometimes the deaths were not self-inflicted: When William Cockburn, professor of philosophy at the University of Edinburgh, died from an "effusion of blood from the lungs," the coroner decided that his death had to do with the juxtaposition of unusually calm weather and a record low in atmospheric pressure.[62]

There quickly followed a medicinal market catering to those who wished to avoid such risk. Newspapers advertised medicines such as Dr Fuller's Lozenges, against "nitrous dampness of the atmosphere"; Balm of Gilead, against cold weather; and Dr Sibly's reanimating Solar Tincture, to be used in the deoxygenized atmosphere. The Nervous Strengthening Pill was praised for invigorating individuals confined to foul air, Marshall's Universal Cerate cured chilblains, Barclay's Candies helped on foggy days, and a few drops of Mr Patence's Diadem Essence were beneficial upon "entering into places where the air is confined and stagnated, such as the House of Lords and Commons, Westminster-Hall, the Opera House, play houses, churches." The powder was hailed as a preventative for "fainting, swooning, shivering, and trembling."[63]

The theoretical concept informing this belief in the correspondence of weather and health was that of *sympathy*. Sympathy explained clinical, physiological, meteorological, and seasonal statistics concerning the atmospheric distribution of health. Sympathy between the nerves and the air could be seen in everyday pathologies: Cold water thrown on a part of a warm body caused a contraction of the vessels and stopped hemorrhages; skating in frosty weather built up the appetite; strong smells instilled vigor in the healthy and caused convulsions in the delicate. The sympathetic theorists blurred the boundaries between the physiological and the aesthetic, the psychological and the climatological; the individual could possess "sensibility" to abstract ideas and heart-wrenching sights, but also to thermal shocks and hard work. For Whytt, this was the case "because the changes in the body occasioned by the sympathy of the parts are stopped by whatever affects the nervous system so strongly, as to overcome the sensations that produced those changes."[64]

An insightful account of sympathy appeared in 1772, under the title *Commentaries on the Principles and Practice of Physik*. The book was written by Whytt's student James Makittrick Adair (1728–1802), who practiced in Antigua, Andover, and Guildford. Adair initially wrote on subjects as disparate as yellow fever and slavery, but eventually settled on physiology and regimen, building his reputation with the *Natural History of Body and Mind*, published in 1787. Moving to Bath, he became an expert on "fashionable diseases." In *Commentaries*, Adair described sympathy as the consent between the "incumbent atmosphere" and morbid changes inside the organism. He proposed that nervous sympathy enabled external stimulation to move into the interior of the body, where the signals continued to travel from organ to organ, until the whole organism was affected. The cold feeling experienced on the skin during windy weather, for example, moved to the heart and intestines, and made them "disagreeably stimulated in consequence of their sympathy with the nerves of the skin."[65] Such transmissions were by no means limited to atmospheric influences, nor were they identifiable by outward appearances only. Adair argued that the effects of changes in weather from settled to stormy could not be distinguished from the effects of shame, terror, or anger. Chills, blushing, flushing, and temporary paleness were caused neither by the mind nor "external causes" solely.[66] Such appearances indicated a fragility of frame, unable to "bear changes of seasons" in a world in which the "employments of life require our being frequently exposed to changes of weather, season and climate."[67]

In the statements like this, Adair was following medical interest in the importance of "predisposing" causes of disease, which, despite their low intensity, produced lasting effects on health. Adair was able to clarify what Arbuthnot and Cheyne intuited: "[T]here are a variety of causes that by their constant, but slow action, weaken the cohesion and spring of the fibres. Such are the moist gross warm air, rainy weather, southerly winds, low damp houses, a situation near marshes, and stagnant waters, warm clothing, and lying too warm at night."[68] The juxtaposition of so many diverse mundane phenomena was underwritten by the sympathetic theory, predicated on the representation of physical space as one of permanent risk. Medical emphasis naturally moved from curing the individual to safeguarding the collective. In this shift, the role of accidental exposure took on a central role.

Among Whytt's students who paid attention to this "accidental" aspect of health risks, John Leake stands as a good example. Having initially studied in Edinburgh, Leake obtained an MD from Rheims in 1763. He became a licentiate of the Royal College of Physicians and started a midwifery practice in London. There he was involved in the 1767 foundation of a hospital for lying-in women, which became his research and teaching ground. The same year, he published *A Course of Lectures on the Theory and Practice of Midwifery,* and ten years later, *Medical Instructions Towards the Prevention and Cure of Chronic Diseases Peculiar to Women* (1777).[69] The latter work went through seven editions by 1792 and became one of the definitive works on female health. Among other topics, it featured a forty-page section on the "Effects of Climates or Sudden Changes of Weather on Delicate Constitutions." Here, Leake drew attention to the inconstancy of health and the variability of bodily functions, which he thought derived almost entirely from accidental exposure to the elements. Such exposure rendered a person "very different at different times, in thinking, speaking and acting; as any one who is not robustly insensible must naturally discover from his own feelings."[70] The majority of "popular diseases" had their origins in the atmospheric agency, whose long-term effects undermined the self-sufficient nature of internal immunity: "Men should not therefore pertly presume on the superior excellence of their bodily, or mental faculties, did they only consider how much accidentally they came by them."[71]

Leake was drawing on Whytt's and Gregory's work. He argued that the observed pathological response of delicate constitutions ought to be taken as empirical proof of the sympathetic effects of external stimuli on the human body more generally. Delicate frames only magnified these influences and made visible that which healthy individuals ignored. Yet Leake also moved a step ahead of traditional accounts in arguing that delicate individuals were mistaken in projecting their feelings onto their physical surroundings. Stressful weather was no more than an "imbalance between the internal and external air," so those who felt dull and heavy erred in thinking that such qualities applied to the weather. Rather, the imbalance "produced a sense of weight and oppression in the body so that we fall into an error, by applying that to the air, which is only the result of our own feelings, from its undue external pressure."[72] Leake sought to replace subjectivity with etiology, but continued to take subjective judgment as the authority in demonstrating the reality of aerial pathology. Whatever delicate individuals felt in rough weather "will more powerfully confirm the reality of such effects than all the reasoning in the world."[73] Leake asked those of a higher nervous sensibility to do their best to keep a parallel record of their health and the state of weather. He asked them to think about how they felt before and after the storm, and whether they grew depressed. Drawing on his hospital experience, he noted that during gloomy weather women became breathless, "like fishes out of water," seized with headaches, vomiting, or nosebleeds, and struck by an odd body tremor. In "oppressive" weather, his patients felt like bloated bags, attacked by pinching bowels and diarrhea, all of which stemmed from the "uncommon stress" laid on the vascular and nervous system.[74] Such helplessness was painful to even contemplate:

> in this frail and *transient* state, the human system is subject to an *inclement* atmosphere without, and the *violent Passions* within; it may suffer from intemperance, and advance of age, and prevalence of *injurious habits*, so as to render it more instable than the *Weather Glass* and in perpetual state of change, from the cradle to the grave. [Leake's italics]...Well may human life, thus surrounded and assailed by a train of unavoidable calamities, be compared to a fleeting shadow which never continues in one stay: like the unballasted bark in a troubled ocean, it becomes the sport of winds and tides, and without the air of Religion, Reason, and Philosophy, is in continuing danger of being swallowed up and lost.[75]

With assertions like these, concerns over atmospheric disease outgrew purely medical concerns, and entered a wider social ambit. Susceptibility to exterior influences extended to mesmerism, somnambulism, and clairvoyance, which, in Simon Schaffer's terms, were turning souls into "marionettes under hidden power."[76] In the medical context, where Cheyne and Arbuthnot made personal responsibility the basis of all attempts to manage weather dependence, Leake's portrayals exuded a disheartening fatalism. The transience of health was the norm; psychological moods eluding self-control and moral accountability were moot issues. Vulnerability to outside agency was beyond individual sovereignty, and free will, by consequence, was in need of a new authority in diagnosis, prognosis, and therapy. The medically defined risks problematized the jurisdiction of human will and challenged the purpose of discipline in hygiene and regimen. In a world in which even *physicians* thought that the state of one's health escaped one's control, one could neither expect to be responsible for becoming ill nor for becoming healthy: "Human life being exposed to many thousands accidents, and its end being hastened by a prodigious diversity of means, there is no care which we can take of ourselves, in any one respect, that will be our preservative."[77] In this context, the insights of Anthony Willich acknowledged that weather symptoms became the most salient indication of the decay of their runaway civilization.

Anthony Florian Madinger Willich (fl. 1799) was a Scot who spent his early career in Konigsberg as a part-time teaching assistant to Immanuel Kant. On returning to London in the 1790s, he became physician to the Saxon ambassador and began to make a living by itinerant lecturing, medical publishing, and writing popular digests of transcendental philosophy. In 1799, in an extended discussion on the evils of sedentariness in his *Lectures on Diet and Regimen*, Willich made an extraordinary claim that, next to gout, the "still more general malady of the times, is an extreme sensibility to every change of the atmosphere; or, rather, constantly sensible relation to its influence" (see figure 2). He noted with astonishment the capacity of some of his patients to identify the direction of the wind even when

Figure 2 J. Caldwell (after J. Collet), *Ladies' Disaster* (1771). Wellcome Trust Image Collection.

inside their apartments. This trait, earlier associated with delicate and ailing individuals,[78] became, in Willich's opinion, a trend among all ranks of people. This was a "talent so peculiar to our age," acquired in the most capricious of European climates, which defined English health in general as "dependent, frail and transitory." Willich went so far as to question whether bouts of political turbulence were due to people's "secret dependence on the weather," because "beings so organized cannot warrant, for a single hour, their state of health, their good-humour, or their physical existence."[79] Modern men and women were creatures of circumstance, not free will.

Contrary to such assumptions, the morally astute demanded that society should not forfeit its accountability under the pressures of brute nature. Traditionalists censured the hypocrisy of aerophobic affectation as mere escapism. They opposed necessitarian views that circumstances fixed the character to the extent that freedom of will looked virtually redundant. Necessitarianism led towards fatalism, destroying "all our Notions of social and political Good and Evil, and is in direct Opposition to every rational System of natural and moral Philosophy." It was provocative to assume, as Hobbes was imputed to have done, that "neither Man nor any thing else took its Beginning from its self; [and that] therefore Men's Actions take their Beginning from something *without* himself"[80]: A materialist, fatalist or a necessitarian were but "different Terms for what is properly called an *Atheist*."[81] Where traditionally people linked moral acts "to something within ourselves, 'tis too true it is all external; the mind rises and falls, quickens or stagnates, just as the Operation of the Powers without direct or relieve it."[82] It was possible to be ill by nobody's fault[83] and to use weather symptoms as a new social excuse. As a contributor to the *London Magazine* put it in 1747: "My health, I acknowledge, is very precarious, and depends so much upon the weather that I am not sure of being in the same frame for two hours together; and whenever I am out of humour, occasioned by the sickliness of the day, I always discharge my company, or absolutely refuse to see them."[84]

Unwell and Exposed

During the eighteenth century, weather-induced discomforts served as a control button for social accessibility, in the same way that nineteenth-century migraines were the stereotypical sexual excuse. This indisposition was often described as "unwellness," a phenomenon that, despite its generality, penetrated all levels of eighteenth-century

middle-class society and its medical culture. The rise of unwellness as
a medical and popular category is an underestimated phenomenon in
the history of the period, and something that presents a new development within Hippocratism and the physiology of nonnaturals. Willich
himself drew attention to the fact that commercial society opened
the door to "a certain weakness and indisposition, whether real or
imaginary," which has "infested society in the character of a gentler
epidemic." He did not think it could be called a disease, but rather
"an approximation to an infirm state" that compelled "man to reflect
upon the *relative situation* of his physical nature, to acquire correct
ideas on health, disease, and the means of prevention of relief and
thus imperceptibly to become his *own guide*."[85] Willich's emphasis on
the relative situation and on self-guidance was crucial. Health management ultimately depended on "positioning" the individual in the
space at once geographical, climatological, architectural, and social.

Unwellness defined the epoch of elegance. The poet George
Crabbe claimed that Lord Chesterfield was the first to introduce the
term into "the vocabulary of fashion" and use it to mean a state of
mild debility, with consequences for personal mobility and fitness. "I
am, neither well nor ill, but *unwell*," he wrote, referring to a feeling
of someone out of sorts with the world, uncomfortable with his surroundings, and unhappy with the state of the air. Chesterfield wrote
to his son that the rain in Germany during August 1758 did not
agree with him, "for it hinders me from taking my necessary exercise,
and makes me very unwell."[86] By the end of the eighteenth century,
the term was in full literary swing. Novelists often used it to capture
the affection of the delicate classes who, as in the case of a fictional
nobleman in the 1794 novel *Angeline*, complained of being "altogether unwell—this air."[87] Physicians blamed associated unwellness
with mental wear and tear resulting in a depression of spirits.[88] It
was manifest in a change in looks, pale skin, or fatigue. At a party
described in *Augusta*, Lord Euston warns his female companion: "I
fear you are unwell. Your fatigue yesterday, together with the agitation you suffered for the distress of your friends, was too much for
your gentle spirits."[89]

In medical vocabulary, the condition was understood to describe
an unease triggered by a changing balance between the sufferer and
his meteorological, mental, and dietary circumstances. The term
could mean a general unhealthiness, but it could also suggest a state
of indefinable discomfort associated with a disturbed "ergonomics"
of daily life. In this latter sense the word is particularly interesting, as
it approximated the "subclinical" symptoms that eighteenth-century

medics typically attributed to the influence of the external world on the human frame: headaches, fatigue, chilliness, but also "self-invited starts, jerks, or twitches" and "tingleing, gnawing restlessness."[90] In at least one discussion from 1785, written by a London expert on pulmonary ailments, unwellness was related to Whytt's "sentient principle," and was presumed to develop in the diseases caused by an affliction of the fiber. The author, Thomas Reid, explained that the working of this affliction was obvious among people with a good health record, who in their old age grew "unwell, without any particular disease, peevish and discontented with every thing about them, restless and moving from place to place." In extreme cases, they would become indifferent to their own condition and die with stoic apathy, "without a consciousness of pain and anxiety." Reid thought that such a sequence could never follow the sympathetic laws between the fiber and the mind, but was instead a testimony to the suffering of "an emanation from the great Creator of the universe, pre-existing upon materials so frail and perishable."[91]

Indeed, the indeterminate quality of unwellness was the key to its popularity: Feeling out of sorts was an excuse to avoid social contact, refrain from exercise, elicit pity, or obtain professional help. In this sense, "feeling unwell" was the point of departure for a larger social pathology, related to an obsessive concern with "environmental sensibility" that took hold of the affluent ranks of eighteenth-century British society. Weather-induced unwellness defined daily routines, not only as something to be managed by regimen as recommended by Cheyne, but also and especially as something that invoked outside stress to shape behavior and cultural aspiration. Unwellness and delicacy provided the nexus in which the bourgeois cult of "feeling" created the first, and prototypical, meaning of "environmental" quality of life.

Speaking of "environmental quality of life," however, does not help us understand the ways in which contemporaries thought about the above issues. Eighteenth-century medical theorists never used the term "environment" as a unifying concept. In modern usage, we are accustomed to this notion as referring to the space, things, and processes understood to be of *relevance* to those who are affected by such space, things, and processes. The environment does not simply exist; it exists to the extent that it bears upon those who find themselves surrounded by it and—importantly—who are exposed to its influences. Eighteenth-century medics did not speak about the environment, but they did speak about exposure to "external influences." The notion of exposure had thus historically underlain that of the "environment." Moreover, it had defined the medicalization of physical space that resulted in what today we call the "environment."

In this last section, I want to connect eighteenth-century medical references to *exposure* with discussions on the socially—and externally—induced vulnerability of the delicate. I argue that contemporary references to exposure and its derivatives were contingent upon the idea of health as a social asset that could be augmented, protected, and dissipated. I also suggest that, in seeking to establish their social identity, the affluent ranks of eighteenth-century British society embraced and perpetuated a *physiological*—if not a pathological—stereotype of hypersensivity to external influences. They did so with the aid of the physiology of the nervous and sensitive man who was endowed with an extraordinary *susceptibility* to the contingencies of the outside world. When the delicate body, as a result of exposure to such contingencies, suffered from unwellness, discomfort, and pain, it required a regimen to reestablish its harmony with its surroundings. In resorting to regimen, cleanliness, and self-regulation, the middle classes seized on routines that ratified their social uniqueness and cultural sovereignty through the emblematic use of "environmental" delicacy. I will argue in the following chapters that the outcome of this differentiation was an outlook characterized by widespread concern over the consequences of the accidental *positioning* of the body in space and time, *and* the pathogenic quality of matter exercising its influence on the body through varied levels of *exposure*.[92] The framing of modern ideas about the risks posed by weather took place through the projection of middle-class sensibilities into a space characterized by ever-increasing exposure.

The notion of exposure to external elements provided an organizing principle to those struggling to make sense of the effects of new regimes of labor, lifestyles, leisure, travel, and everyday behavior related to ambiental health and the upkeep of basic health. "Exposure" also made sense in analyses of regional susceptibility to the geophysical features of land and water in relation to endemic disorders. But exposure is helpful especially in understanding the medical construction of sedentary, industrial, agricultural, and military pursuits, as far as they were understood as defining an "objective" vulnerability to exterior risks. In this sense, the term had a quasi-technical meaning, used to described the fact that "in all employments of life there will be an exposure to the causes which produce disease," whether to the open air, in agricultural occupations, or to the atmospheres of mines and manufactories.[93] Becoming sick, on a general level, had to do with the fact that the body possessed only so much inborn ability to ward off the inconveniences of daily life, which, in the words of doctor John Cheshire, "Man, from the nature of his Existence is continually expos'd to; the numerous Inclemencies of the Air, the Irregularity in Eating or Drinking, or in any of the other Non-naturals."[94]

While its use dates back at least to Sydenham (the patient was told not to "inadvisably expose himself to the cold air"), eighteenth-century references to exposure included those to "the air," "heavy rains," "inclement sky," "moisture," "winds," "smoke," and "sun beams." But they also included exposure to "hardships of fatigue," "voyeuristic eye," "insults," "despotism," "torture," and "the enemy shots."[95] The term referred to the predicament of being in a condition or place subject to external hazards, regardless of whether these hazards originated in the natural or social world. In the medical context, exposure had a negative connotation, because it referred to the circumstance of "being exposed" to the exciting causes of disease: "fencing, tennis, hunting, and other diversions require much greater exertions, and expose also to the severest injuries of the air."[96] On other occasions, such causes were connected to local climates, in which, for example, disease came about "from a long exposure to an unusually moist state of the atmosphere."[97] In some cases, hazards could be brief and sporadic: "the causes of [aphthous sore throat] seem to be exposure to cold air [...] children, particularly boys, who use violent exercise in hot weather, and soon after expose themselves to a current of air or drink cold water are most liable to be affected with this species of Miliaria."[98] In all such references, however, medical writers assumed the existence of a dichotomy between the passive and active elements: The exposed person passively received the "influence" of an active agency.

The term was extensively used in other contexts. Exposure in chemistry signified the action of one substance on another; in architecture, it referred to the position of a house with regard to prevailing winds and lighting. The law defined vagabonds as "endeavoring by the exposure of wounds or deformities to obtain or gather alms."[99] In the language of finance, "commercial property is exposed to so many unavoidable hazards" that it was necessary for the laws "to induce people to insure the risque."[100] In politics, "exposing" somebody's errors, lies, and secret plans indicated public embarrassment, fall from grace, or an outright scandal.[101] In literature, it referred to a state of impropriety and nakedness. In John Hemet's *Contradictions* (1799), a character has to run an errand in a shower of rain, but is loath to leave his female friend without an umbrella and "expose [her] to all the inclemencies of the weather."[102] Whether an umbrella, hat, or roof, the existence of a shelter was implied in all usages referring to the unpleasantness of a space outside one's intimate comfort. "In our own, or our friend's drawing room, we are so much *at home*, that any avoidance of draught, or precaution against exposure, is unthought-of."[103] In such pronouncements, polite domesticity contrasted with outdoor drabness; a gentleman who changed his rain-drenched clothes ceased to look like a "weather-beaten traveler," but appeared to be "the elegant man of fashion and the world."[104] Concerns over risks, health and civic virtue could be conjoined to indicate an action that triggered a train of pathologies, both social and medical, as was the case with the problematics of "exposed and deserted children" cared for in foundling hospitals.[105]

The fusion of somatic and social exposures signaled a major moment in the history of British and European medical theory and practice. A contained problematic, addressing the conditions of a delicate population, now became a universal indicator of environmental life, marked by the medically biased notions of sensibility and vulnerability. The curse of the new generations lay not only in what they could (but failed) to rectify (such as *bad* air), but also in what they could never hope to escape (*air* itself). "Air was the one of non-naturals affecting the bodies every moment of our lives and from which there was not escape, for good or ill."[106] And, as shown above, crucial in bringing about the ubiquity of exposure as a condition of modern life were both the vitalist and mechanical readings of the air's role in physiology. In their stress on the total interpenetration of the organic and the nonorganic, these readings made it possible to see disease as a systematic, ongoing process, rather than an accidental event. Whether as a finely tuned hydraulic machine or a strung-up network of nerves, eighteenth-century physicians represented the body as an entity under constant pressure to yield its health to outside powers. In addition to the random risks of contagion or miasmatic infection, disease was a perpetual possibility, to which a methodical regimen was the only appropriate remedy.

Surrendering jurisdiction over one's health to a force outside one's control in this way might have been as disconcerting as having already fallen ill, but it was no worse than surrendering one's reputation to gossip and so ceding control over privacy and peace of mind. Both social and medical exposure compromised valuable assets—privacy and health. Both were seen to represent hazards to unalienable medical and civic rights. And where social exposures meant embarrassment and scandal, atmospheric exposures compromised health. More specifically, medical studies of the air inaugurated a patient whose condition depended on contexts, not immediate pathological changes: a person was not sick on account of her humoral imbalance, but because she had problems coping with heat.[107] Eighteenth-century physiology thus gave a basis to commentary on the fashionable diseases that, according to the medical men, were taking their toll on the population at the epicenters of British consumerism and entertainment. This view, I stress, informed the control of the individual and social conditions of existence, which I will shortly discuss in relation to dress, domestic appliances, architecture, and health travel.

Chapter 2

Cursed by Comfort

Air being one of the most necessary things towards the subsistence and health of all animals, 'tis a wonder to me, that here in England, where Luxury, and all the arts of living well, are cultivated even to a vice, the choice of air should be so little considered.

George Cheyne[1]

In time, this pure air may become a fashionable article in luxury (although) hitherto only two mice and myself have had the privilege of breathing it.

Joseph Priestley[2]

Atmosphere Enclosed

Crucial to the eighteenth-century analyses of everyday disease was the role of the indoors, which has so far received only a marginal treatment in environmental history, and still less in the history of science and medicine. No studies exist on the pathologies of bathroom humidity, church chills, basement dews, kitchen smells, and bedroom heats. Historical accounts of environmental medicine proceed on the assumption of weather as an "outside" entity evolving away from shelter, comfort, and warmth of the domestic realm. The weather is seen as outside the physical spaces in which people live, work, sleep, and entertain themselves for at least eighty percent of their lives. Yet despite such perception, the weather would hardly merit notice were it not perceived from somewhere else; i.e., from within the space of domestic ease, from the safety of shelter and behind the window of a weather-tight house. Such interiors possess meteorologies of their own that were far from insignificant in the daily maintenance of health and comfort. In fact, limited volumes of indoor life were precisely those about which people worried most and which they even framed as elements of their intimate climates. It is perhaps the commonplace nature of such intimate climates that led to their neglect even if they were known to exert far from trivial effects on the medical quality of everyday life.[3] In reality, it was often the case that culinary smells, coal smoke, ill-fitted windows, "effete" bedroom air, fireplace drafts, candle stench, and soggy corridors spurred medical concerns and inspired the first improvements in domestic hygiene, sartorial practices, public health engineering, and architectural intervention.[4]

By the late eighteenth century, physicians shared a belief that many disorders resulted from excessive exposure to the ills of indoor life. The diagnosis coincided with the eighteenth-century ascent of "public sphere" that spurred enlightened learning and sociability in urban settings, especially in places like coffeehouses, circulating libraries, learned societies, and lecturing theaters. Although such spaces shaped the intellectual and political ferment of early modernity, little attention has been given to how these spaces affected the *health* of the new public.[5] Habermasian public spaces were extraordinary environmental singularities in which a juxtaposition of cerebral enlightenment and visceral effluvia produced some of the most breathtaking achievements of the period. Enlightened pathologies included claustrophobia, chronic fatigue, dizziness, drowsiness, hyperventilation, irritability, and mental disorders of any description. The medical intervention that attempted to address such evils involved the sanitary, architectural, and comfort technologies for improvements in heating, ventilation, lighting, window construction, and house layouts. Such interventions had precedents in research into hygienic conditions of special places and "closed institutions" such as prisons, workhouses, and ships. Domestic space was left out on the grounds of its ordinariness and low density. But with unprecedented demographic squeezing and overcrowding, it became apparent that they too required medical attention.[6]

Losing sight of such spaces might be understandable from both historical and contemporary points of view. Most indoor nuisances were minor and easy to cope with. Most were benign. But conditions such as confined air, lack of good plumbing, smoke, humidity, pollution, and overheating were seen as increasingly intolerable. They were represented as a result of social and economic trends with major implications for the identity of the nation as a whole. Commentators argued that many such conditions stemmed from a change in material circumstances of urban living and from the exposure to "unnatural" levels of work or lifestyle excess. Some conjectured that the spread of typhus owed to "the *unnatural* state in which the inhabitants of manufacturing towns are placed."[7] Others considered the possibility that the sheer "artificial constraint and confinement in narrow space" induced an abnormal secretion and caused contagious fevers in jails, ships, and hospitals. "It indeed appears, that the accession of pure air, and the active employment of the limbs or powers of motion, are the principles given by the Author of Nature to preserve the health of the animal system; for whenever the human body becomes deprived of these essentials, its health languishes, and its vigour decays."[8] Typhus and contagious diseases were divine retributions for the overstretching of natural powers. Although commercial activities might have been invigorating for those who pursued them, excessive industry, for which the human body was *not intended* and into which it was steeped by artificial wants or avarice, made the organism retentive of poisons and prone to infection. James Johnson, perhaps the most influential writer on medical meteorology in the early nineteenth century, thought that the intellectuals, bons vivants and businessmen thwarted their energies by exposures to stress, deadlines, and the lack of exercise. "Especially in this country," Johnson observed, "and in consequence of the immense interest in politics, religion, commerce, literature, [and] a stress due to speculative risks [...] it is astonishing to observe the deleterious influences of the mental perturbations on the functions of the corporeal fabric."[9] Johnson, like many before him, thought that main cause of decay lay in the ascent of *luxury*.

Luxury provided a master narrative for medical thinking about the everyday quality of urban life, but its role in contemporary culture extended far beyond its medical meanings.[10] Together with corruption, anarchy, and rebellion, luxury was one of the most debated themes in the literary and political commentary about eighteenth-century British mores. Ritual condemnation of its influence on minds and bodies became a catchphrase in the ideological battle to legitimate aristocratic rule and define culture in terms of masculinity, temperance, and moral integrity. Regardless of political inclination, the public spirit was often portrayed as nearly equivalent to abstinence from luxury. The elimination of "the moral and natural causes" of political degradation became more important than any specifically "political" act of amelioration. In the shared consensus over the dangers of luxury and effeminacy to which it led, temperance, modesty, and rationality were emblematic of the English national character. "Temperance and patriotism go hand in hand," wrote Nathaniel Lancaster in 1745, suggesting that the tendency to "squander away immense sums upon foreigners and barter the sinews of war for an effeminizing delicacy [...] is perhaps as unsuitable to our genius and climate, as it is inconsistent with the care and solicitude which danger awakens in the patriot's breast."[11]

Eighteenth-century political theorists sometimes argued that prevention of political decline by good manners was more effective than formal acts of legislation. Acting preventatively and thinking circumstantially had more power than acting reactively and thinking bureaucratically. Such expectations were not limited to ideals of political virtue and social order: Prevention and circumstance were the keys to the ideals of somatic virtue as well. As political rationality and social demeanor became matters of characterology—measured in terms of virtue, education, compassion—so too the thinking about health and diseases became embroiled in thinking about character, conduct, and bodily care. In such context, luxury was seen as undermining the character as much as the health. Luxury bred a weak personality and tender body; it engendered "effeminacy," hypochondriasis, and the culture of complaint. It spoiled. And it gave rise to a tendency to absolve personal flaws by invoking a susceptibility to influences over which a victim claimed to have had no control, from which he or she could never fully escape, and that could be identified as the real causes of his or her personal mistakes. Susceptibility to meteorological conditions was an example of this tendency. Weather moods, as Samuel Johnson put it, were chimeras "operating on luxury" and it behooved the virtuous man to defy "the tyranny of the climate, and refuse to enslave his virtue or his reason to the most variable of all variations, the changes of the weather."[12] Tobias Smollet, himself an indefatigable critic of luxury, referred in his *Humprhy Clinker* to the mind of those whose "opinion of mankind, like mercury in the thermometer, rises and falls according to the variations of the weather."[13] But it wasn't the weather that caused fickleness as much as the fashion of living above one's capacities. Smollet captured the effects of this transgression in describing the atmosphere of London's entertainment, in which "I breathe the steams of endless putrefaction [that cause] those languid, sallow looks, that distinguish the inhabitants of London from those ruddy swains that lead a country life—I go to bed after midnight, jaded and restless from the dissipations of the day—I start every hour from my sleep, at the horrid noise of the watchmen bawling the hour through every street, and thundering at every door."[14]

Fueled by luxury, city pleasures encouraged sloth, sedentariness, and dissipation.[15] Even in Manchester, doctor Richard Price thought that the town suffered a high mortality for reasons that were "first, the luxury and the irregular modes of life which prevail in towns; and secondly, the foulness of air."[16] In 1776, a health manual advised that children should not be "brought up in great cities or large towns where the air is always polluted by the exhalation of dead animals, or the unwholesome vapors proceeding from standing waters."[17] Many critics raised voice to condemn city demographics while praising rural retreat. By the 1760s, the polarization between urban decay and the purity of the countryside found expression in the literature of agricultural moralists like Adam Dickson, William Harte, and John Mills, who, following in part the French school of *agronomistes,* praised outdoor activity, wholesome diet, and the healthiness of the country air. The peasant archetype became the norm against which the urbanite looked like a travesty of virtue and health. The polarization embodied the worry about the direction in which urbanization and luxury were taking the English nation. In the views of Samuel Johnson's friend and physician Percival Stockdale, the city dwellers required activity to "preserve the health of the animal machine" because indoor life enhanced sensibility through languor and artificial heating.[18] Heat sensitized the skin nerves to further stimulation while sweating made cuticles fitter for the transmission of stimuli that led to putrefaction. The result was a weak, sickly, and obese body that nurtured a tropical psyche, at once passionate and amorous, vindictive, and timid. If this was becoming a widespread phenomenon, then health was to become a matter of national importance, the political implications of which were beginning to find expression in philanthropic arguments developed around midcentury.[19]

On this subject, especially well-articulated were the views of Jonas Hanway (1712–1786), whose activities from around the 1750s bore the mark of an age that was becoming receptive to conceiving the individual's state of health as determined—to a certain extent at least—by an environmental imprint of the collective. Born in Portsmouth in 1712, Hanway spent twelve years as a merchant at the English Factory in Lisbon and more than five years in Saint Petersburg. On his return to London in 1750, he continued a business venture as a Russia Company employee, but gradually devoted himself to supporting associated philanthropies. He was on board of governors and subscription committees for the rest of his life, writing books and pamphlets and working to improve the parish care of pauper infants, even successfully lobbying for parliamentary reforms in this matter. Although his interests ranged from Jewish naturalization to the importance of fresh bread, from the conditions of servants to the health dangers of late hours, the enduring question in all of these studies was whether lives could be saved and people trained to serve the national growth. "You will easily perceive how Health, Population, Industry, Wealth, Comfort and National Felicity, are linked in a Chain with each other."[20] It is in this context that we must look at his fascination with the "artificial" meteorologies of closed space.

In *Serious Considerations*, Hanway viewed these conditions as preventable evils arising from economic circumstances of midcentury Britain. In this work he thus went beyond his recommendation about the care of pauper infants and considered the issue of air quality as tantamount to engaging with the looming decline in national health and economic well-being. *Considerations* is essentially a moral diatribe that, for all its predictable rhetoric, contains insights into what it must have been like to try to redefine common facts such as crowding as medical problems. Hanway ritualistically condemned late-night entertainment when the tipsy crowd gathered in such numbers "as if we thought there could be no Pleasure without being crowded"[21] Hanway quarreled with the practice of shutting the windows in rooms where leaving them open an inch or two would make all the difference. While he admitted to being aware that for many the cold presented a greater danger than stale air, he nevertheless considered it a prejudice leading to chronic debility. If people trusted their own sense of oppression and discomfort, they would avoid dangers of rebreathed air in playhouses, assembly rooms, public sale rooms, and crowded apartments of all description. What was true about the aerial conditions of ships, hospitals, and prisons was, in Hanway's view, equally true of the spaces of usual social intercourse. But Hanway worried that such risks were hard to isolate and make the subject of medical concern. Tainted air was tolerated either because it was unavoidable or because it wasn't perceived as such: "[F]ew Persons are sensible of half their Danger in public and private Places for their being crowded and confined [...] Public places of every kind, even under the meridian sun, are hurtful to some constitutions; but how rarely are people sensible of this! What renders them cheerful, they suppose gives them health: but a crowd is no sooner assembled, than the air becomes unfit for them, whose thread of life is easily cut."[22] The "imperceptibility" of aerial threat remained a nagging concern among medical educators long after Hanway.

Hanway proposed a list of improvements in window construction, existing architectural features, and air circulation methods. While some of these will be discussed in the chapter, it is worth recognizing that his and others' experiences of environmental threats made sense in the context of commercial society and the negative readings of luxury. Invoking the stereotypes of national decline, Hanway challenged the existing ideas about indoor environment. The classic trope was that of "fresh air," which had been a mainstay of health for generations. Sydenham insisted on its prophylactic qualities, and Herman Boerhaave held it in esteem in helping lung obstruction. "A laborious hard life, in a free country Air," with an occasional "following the Plough," were excellent practices for all, not just the sick.[23] Other advised it to calm arrhythmia, reduce gout, stop early consumption, and improve nervous energy.[24] William Falconer, a doctor and scholar from London with reputation in medical meteorology, wished to prevent urban "deskism" and laziness with good doses of field labor in an air impregnated by vegetable effluvia. Outdoors made people robust, whereas being indoors relaxed them into disease. Nervous complaints arose from confined atmospheres "to which the modern style of domestic and social accommodation, necessarily exposed most of the higher ranks of people." The ills of "accommodation" were further compounded by a craving for gratification that led to sloth, indolence, and hypersensibility.[25] Fatigue, broken sleep, uneven pulse, and fevers also arose as consequences of comfort and inactivity.[26]

Whether we look at Hanway's innovative thinking about the sash window or Falconer's preference for outdoor therapy, eighteenth-century physicians worked to conceptualize social morality in medical terms.[27] Phrasing medical risk in a language of moral failing—as we now do for climate risks in general—physicians asserted the right to pass judgment on social trends that they routinely connected to urban lifestyles of the middling ranks. They served images of modern gentlefolk degenerating into overstimulation by "unnatural" stimuli. While sensibility defined propriety in a genteel society, medical practitioners noted its other, more insidious side. The medical picture of a delicate *homme du monde* was one of a sedentary creature immersed in the phlogisticated atmosphere of the cardroom, in which his moral and physical faculties sank to stupor or sin. It is not surprising that such readings often came from the vigorous Scots, who were offended by the ways in which continental effeminacy took hold of English (and increasingly Scottish) metropolitan life.

One such Scots was James Gregory (1753–1826), son of professor John Gregory of the University of Edinburgh. In 1767, Gregory went to Edinburgh to study medicine under William Cullen, Alexander Monro *secundus*, Joseph Black, and John Hope. Following his father's death, James took over his lectures, graduated in June 1774, and spent two years in Leiden and Paris. From 1776, Gregory started a successful Edinburgh career as a professor in the institutes of medicine. He became a leading teacher of the faculty, as well as an accomplished classical scholar, public figure, and practicing physician. Among his many publications, his dissertation, "*De Morbis Coeli Mutatione Medendis*," deserves a special interest for its clarity and original treatment of climatic causes of disease. The dissertation was first published in 1785, but was translated in 1815 as "A Dissertation on the Influence of Change of Climate in Curing Disease." The largest portion of it deals with the pathologies arising from exposure to outdoors and physiological adaptation to seasons, but there is fine thread of thought that gradually emerges to address the social origins of delicacy.

Gregory notes, for example, that while "incursion of climate" might be inevitable, it might also be probable that at critical periods and in certain constitutions, "there are idiosyncrasies arising from peculiarities in the manner of living [...] which render persons who are in other respects perfectly vigorous, liable to diseases incidental to climate."[28] He points out that the arguments about climatic influence may already be worn thin from overuse. Doctors and patients often refer to causes that only superficially may be related to the air. On closer inspection, however, climate may be etiologically inert. This could explain why native climates never hurt local animals: "[U]nder circumstances of the greatest possible exposure to the power of its severity [...] adaptation is often sufficient to stop disease." Divine hand thus prevents any "naturally" caused climatic disease, demonstrated by the fact that humans endure all weather transitions "if they lived temperately." Those accustomed to hard outdoor labor enjoy health like other animals, but once societies developed the means to protect bodies artificially from seasonal extremes, "man was subdued by the severity of his native clime." Those who had to withstand the weather because they could not choose otherwise, "seldom suffered any inconvenience from it; and it is only in proportion as men become *afraid of exposing* themselves to the intemperance of the air, and sedulously endeavour to avoid it—that they are rendered susceptible of its effects." For Gregory, such sheltering was the consequence of acquired inactivity and thermal equality of modern residential life. Delicate people were *made* by overprotection. It was therefore not surprising that those who were used to such conditions, "and to indulgence in various luxuries, should be solicitous to elude the rigorous excesses of the weather."[29]

This was a strange and self-perpetuating feedback loop: The more one avoided outdoors, the more one felt it *ought* to be avoided. The indoor life became both a source and consequence of delicacy and thus an axiom of Georgian social dynamics. In contrast to the investigations of seasonal disorders and atmospheric disease, doctors increasingly recognized the need to look indoors for answers about the failing health of the middle-class populace. James Adair, introduced in the previous chapter, led the way in this direction as a practitioner with ample experience in fashionable diseases acquired during his years of practice in Bath. Adair prefaced many of his medical texts with observations on the effects of fashion and "artificial wants." For example, in his *Medical Cautions* (1786), he explained that the very idea of discussing fashion in a work on medicine bore witness to the fact that medicine itself had came under the influence of the intellectual fashion of talking about conditions that had not been observed before. Fashion, argued Adair, had long influenced the well-to-do in the choice of physicians, but it had not yet become clear how it had influenced them in the choice of their diseases. The latter possibility was predicated on the fact that people of fashion tended to appropriate an "exclusive privilege of having something to complain of" to the point that the sheer act of complaining had introduced "medical" terms—spleen, vapors, hype—that had no referents in reality. In Adair's view, such illusions had cultural and class origins, but they were very often associated with the "topological" conditions of urban life and, more concretely, with a trend that made elite sociability increasingly allocated to residential space in which it had not been seen before.[30]

For Adair, the misuse of residential space was one of the more neglected medical issues of the day. He identified the indoors as a serious cause of (feigned or real) afflictions and did not spare effort to look into how Bath's windiness, street layout, and architectural features affected his patients. He in particular looked into the ways in which the uses of rooms altered the circulation of air and its quality. He found that the combination of fireplaces, candles, and closed windows kept the interior heat well above his standard of fifty-eight degrees Fahrenheit. Normal breathing, home fires, and perspiration destroyed a dozen gallons of air in a minute and had an impact on vital power: [I]n such an air, individuals became languid, short of breath, sweaty, and depressed. The effects worsened with the concoction of smells of wall paint, flowers, perfumes, cleaning products, and kitchen odors. Uninterrupted winter heating increased the warmth of walls, which disturbed sleep, reduced resistance to fevers, and, in many cases, led to acute disorders. Adair recounted a soiree during which a guest fainted during a card game but whose condition only got worse with the doors open: His delicate constitution "was much injured by the sudden exposure to the current of cold air."[31] He was concerned to recognize that in such cases people ascribed the cause to a bodily malfunction when the real cause lay outside. Adair thus chastised female readers for blaming bile for their jaundiced-colored eyes; the real cause was the tainted air around the card table in a room in which crowded lung patients often succumbed to "swoons."[32]

Polite entertainment caused impolite odors. Tobias Smollet addressed the connection in his *Humphry Clinker* when he described a ball party at which, after several hours, stifled amid a raucous throng, Matt Bramble began to wonder how so many "hundreds of those that rank as rational creatures" could decide to bevy in a clammy air and enjoy a dance that caused "a swimming of the head which was also affected by the foul air, circulating through such a number of rotten human bellows." Even at the door, where he presently removed himself, Matt fell ill as if all of a sudden, "came rushing upon me an Egyptian gale" infused with morbific vapors that forced him to drop senseless on the floor. Back to life with the help of Sister Tabby's hartshorn handkerchief, Matt is looked at by a doctor, who explains that his swooning was "entirely occasioned by an accidental impression of fetid effluvia upon nerves of uncommon sensibility." Bramble is vexed: "I know not how other people's nerves are constructed; but one would imagine they must be made of very coarse materials, to stand the shock of such an horrid assault. It was indeed a compound of villainous smells, in which the most violent stinks and the most powerful perfumes contended for the mastery. Imagine yourself a high exalted essence of mingled odours, arising from putrid gums, imposthumated lungs, sour flatulencies, rank arm-pits, sweating feet, running sores and issues, plasters, ointments, and embrocation, hungary water, spirit of lavender, asafoetida drops, musk, hartshorn, and sal volatile: besides a thousand frowzy steams I could not analyze. Such is the atmosphere I have exchanged for the pure, elastic, animating air of the Welsh mountains."[33]

Like Bramble, Adair was puzzled by the social proxemics that dictated such parties: why, despite the spaciousness of Bath's public rooms, did people prefer to cram in bedchambers, closets, and even cupboards, "not only to the injury of the public institutions, but manifestly, to their own health?"[34] Why not redesign the window casements whereby their upper parts would remain open and allow for a continual supply of good air? Why insulate the house against the cold at the expense of air quality? The fact that fresh air flowed into the room under the door to compensate for the outward flow of tainted air was enough to show the mistake of insisting on "double doors, linings, listings and sand-bags." Avoiding such practice would prevent accidents like the one in London's Golden Square where eighteen charity children, to render the room warmer, closed the chimney and crevices to exclude the cold air, which made ten of them subject of major neuralgias, lightheadedness, and convulsions. Adair regretted the lack of knowledge that led to such incidents, but he also queried the enlightened of his well-heeled customers who used private apartments for the purposes and crowds for which they were never constructed. Could this be an example of a "trickle-up" effect on the upper classes discovering the erotic conveniences of small spaces from the experiences of their servants?[35] Whatever the reason, crowding, in Adair's view, was excusable in London, where larger apartments could compensate for the lack of purpose-built card rooms. But in Bath's smaller flats, routs had little justification; they thwarted recovery while helping physicians' earnings. When Adair complained about this to his friend, the latter replied: "Let them alone Doctor, how otherwise could twenty-six physicians subsist in this place?"[36]

Adair was aware that the atmospheric conditions during nocturnal bacchanals had extraordinary properties. This was commonly observed in an age in which any minor or major gathering required lavish illumination. Lady Cowper's party in 1769 was lit up by five dozen wax lights. The Duke and Duchess of Northumberland's gala boasted four glass chandeliers, each with twenty-five beeswax candles. By 1802, Mr. Thomas Hope's "splendid rout," with one thousand guests roaming through sixteen rooms (averaging more than sixty people per room), was illuminated by 250 wax candles, "many exceeding two pounds and a half." According to a later historian of ventilation, these numbers would make a quarter of the rooms' air irrespirable within an hour.[37] The effects of breathing, perfumes, perspiring, and overdressing combined with candle smoke and fireplace glow to produce some of the worst airs known to the Englishmen since the Black Hole of Calcutta, which brought on 123 deaths in the stifling dungeon in the Fort William after it was captured by Siraj ud Daulah.[38]

More work on interior atmospheres came from Anthony Willich, who stressed the importance of domestic scales of risk and its role in decreasing the body's ability to "resist the noxious agency of powers, which affect us from without." Arguing that the susceptibility to weather resulted from the "change from a hardier to a more relaxed state of life," Willich proposed that the future of public hygiene depended on an ability to control atmospheric qualities of residential space in a way that would prevent further detachment of bodies from

Figure 3 J. Barlow, after S. Collings (1790). Men sitting at tables smoking and having a merry time; one man has fallen to the floor and spilled his tankard of ale. Wellcome Trust Image Collection.

outside conditions. In their room-centered analyses of medical risks, Adair and Willich could be described as the most prominent "cameral climatologists" of modern medicine.[39] Willich divided room air in three layers: the lower region of carbonic gas, the middle part of a better-quality air, and the uppermost layer of inflammable gas produced by human respiration and fuel combustion. The idea was to expand the healthy middle portion as high as possible. This might be achieved by raising the height of ceilings to keep the lower boundary of bad air above the inhabitants. In addition, all rooms could be aired or "pumped" by brisk moving of the door backward and forward. House residents were advised to administer small explosions of gunpowder to reduce clamminess of the air. On a dry summer day, they could use vinegar sprinkling to increase humidity. Or they could use plants to increase oxygen levels while getting rid of superfluous candles. Equally risky were exhalations in bedrooms containing green fruits and wet linen.[40]

Hygienists also worried about the long-term effects of indoor smoke, as well as the hazards connected with technologies for its elimination.[41] Doctor Robert Bath wrote on the need to adapt heating to seasons: "Early fires in autumn are particularly necessary; as we are to consider that the occasional damp, and vapour arising from frequent rainy and moist days hath never been rarefied and done away by fire, in the summer season." Moist air condensing on the wallpaper, wainscot, and plaster rooms could cause colds especially among those with a "mixt residence in town and country." Danger spread in both directions: Those moving to spend the quieter part of the season in the country moved into the unaired and mildewed quarters in need of being "broken-in" to comfort by use. On the other hand, those taking a job in London ended up among "damp walls, damp floors, and frequently new painted rooms not sufficiently exposed to the sun or dry air." Further fears were raised about cleaning practices that dangerously increased humidity and about the practice of moving rashly into newly built houses even after they had been dried by fires. Because moisture extremes accounted for outbursts of most autumnal fevers, rheumatic pains, and ophthalmias, Bath reasoned that the prevention had to be formulated in terms of comfort in dry air and the "habit of body as near as possible to the same sensations." Seasonal changes could be excluded and kept on the outside by means of the attentive thermoregulation of living space.[42]

This, of course, was a recipe for damaging immunity. Doctors who analyzed the seasonal effects on health had noted that prolonged indoor life during the cold part of the year rendered many unwilling to face the approaching warmth and sunshine. The observant medical pneumat Thomas Beddoes explained that in early spring, "I know not if one hears more murmurs against the sun from any lips [...] The smallest movement tires; fine weather relaxes, the duties and civilities of life are too much to go through." Beddoes, who was experienced in fashionable ailments, had given considerable thought on how lifestyle connected with weather complaint.[43] He called a "law of sensibility" the condition in which "the susceptibility of impression increases faster than ingenuity can bar out external agents. [I]n the best secured fortress of effeminacy, it is the fate of the occupant to shiver more at the inclemencies of the seasons that the mountaineer who is exposed to all the blasts of winter"[44] No amount of care and protection would suffice to eliminate external stress. If anything, domestic ease would only increase vulnerability.[45] The self-imposed fragility of urban middle classes projected a deep-seated fear from a loss of control over one's conditions of life. People didn't like to be discomforted anymore than they liked to be caught in a spring shower. However seemingly rational, such reasoning was seen to undermine health, as the popular science writer George Adams put it in his *Lectures on Natural Philosophy*: "[T]he luxury and effeminacy of this age is studious to stop up every crack, and exclude, as much as may be, every breath of air, wrongly consulting present indulgence, at the expense of future ease and comfort [...] if you are wise, you will be forward to expose yourself to the freshness of a cool air."[46]

Such warnings vindicated those scoffing at the effeminacy of city life against the benefits of bracing country air, exercise, travel, light clothing, and unfettered exposure to the elements. Some made it a creed to live as "naturally" as they could, and their heroic attempts at physical endurance bolstered the idea that health equaled Edenic purity even if in the idea was a thinly veiled nostalgia for a lost world. In reality, even "naturism" relied on reform in domestic and urban space. The advice was to place beds underneath open windows, use heating sparingly, let light in, provide ventilation, dress lightly while indoors, and practice morning "air baths." Uplifting examples of rejuvenation promoted somatic stoicism, such as the apocryphal account of John Wesley's hardships during which "he exposed himself with the utmost indifference to every change of season, and to all kinds of weather, snow and hail, storm and tempest. He frequently slept on the ground in the summer, under the heavy dews of the night; and in the winter with his hair and clothes frozen to the earth in the morning."[47] Wesley could have been outdone only by James Graham, famous for his three hundred therapeutic self-burials in wet soil that he carried out regardless of type of soil and severity of the season. He claimed to have slept covered with turf at night, and in early 1793, in an ultimate feat of defiance, he stopped eating entirely, applied clods of earth to his breast, stomach, and bowels, and embarked on a two-week diet made up only of water and air. In an affidavit sworn on April 3, 1793, Graham declared that "by these means alone, without any other food or drink, I had sufficient strength and spirits to go through the daily excessive fatigue of great and extensive medical practice." Earth, water, and air were the "all-animating material breath of God" that could, in the troubled times of international politics, become a survival kit in a besieged town without supplies. Graham finished his diet seventeen days before France declared war on the British.[48]

Aerophobia

Even for a public accustomed to the panacea market of the late eighteenth century, living on air or sleeping under turf would not have been typical methods of ensuring a long life. In the city, health took back seat to pursuits of pleasure. This might have been true especially in England, of which the world thought was the most advanced in the procurement of comfort in domestic space: "[I]n every thing, indeed, which appertains domestic economy, the English people may be said to excel."[49] But sociability and domestic economy also meant more time indoors: In late Georgian England, this time could amount to four-fifths of a life, a good part of which, depending on household or employment, ran its course in an absence of fresh air.[50] If such an air might have been tolerable to "valetudinarians in reputation as well as constitution [who] avoid the least breath of air,"[51] it chipped away at the stamina of the healthy. By the end of the century, physicians put extraordinary emphasis on what daily exposures to environmental stress in urban setting meant to popular health and medical practice. In this way they fashioned, in Barbara Duden's view, "the hierarchy of illness-causing phenomena [...] as part of the logic of a life story, not as a part of the logic of the body as such."[52] Health was neither innate nor intrinsic to the body but equated with the circumstances *outside* the body that related to one's situation, education, and rank. This was an animistic view that projected organic properties into inanimate space and made health and comfort a consequence of living in "healthy" houses, "healthy" countries, or "healthy" climates.

But this also meant that discomfort and unwellness depended on the lack of care, cleanliness, and convenience in the domestic realm. Indeed, the late eighteenth century could be seen as the time when complaints of discomfort took on epidemic proportions. Feeling uncomfortable or "incommoded" had less to do with a specific condition than with a culture of complaint in which such epithets betrayed the everyday of urban living and unmet expectations. Discomfort was associated with "carelessness and dirt,"[53] and equated with misery, cold, improper clothing, and bad climates.[54] It was mentioned alongside unwellness, hypersensitivity, and hypochondria. Thomas Beddoes prefaced his *Guide for Self Preservation* with a wish to help mitigate "those lighter ailments, which, in this fickle climate, often render the half life uncomfortable."[55] Satirists took up the trope to connect it with Macarone's effeminacy, absurdities of dress, inconveniences of domestic technologies, and newfangled mores of affluent upstarts. Artists such as James Gillray and Thomas Rowlandson made series of prints on "miseries" and "comforts" of anything from traveling to matrimony. In doing so, they made an acerbic counterpoint to the optimism surrounding the new household appliances like the Rumford stove, umbrella, armchair, and gout prosthetics. In his study on the history of design and the invention of Anglo-American comfort, John E. Crowley showed how this criticism responded to the fact that "comfort" represented a new development, not a response to a "natural" demand for higher standards of living.[56]

Criticism and satire centered on the inconveniences in the mundane microenvironment. This was documented in the hilarious *The Miseries of Human Life* (1806), written by James Bereford, which ran to eleven editions within a single year to become an instant classic. The book is a compilation of complaints made by two unusually perceptive characters, Timothy Testy and Samuel Sensitive, whom Bereford endowed with a capacity to detect the minutest faults associated with life in London, entertainment, traveling, reading, and the distresses of medical nature. In describing the "miseries," Bereford focused on minor sufferings of everyday quality, because, in his view, they took precedence over the debacles of nature such as earthquakes and epidemics. Daily irritations caused more grief than rare disasters because the latter had a place in a theodicy of the "greater good." Unlike disasters, however, mundane discomforts had no theodicy: This made them hurt even more. What are the "*Senses* but five yawning inlets to hourly and momentarily molestations? what is your *House*, while you are in it, but a prison filled with nests of little reptiles—of insects-annoyances—which torment you the more, because they cannot kill you? and what is the same house, when you are out of it, but a shelter, out of reach, from the hostilities of the skies? What is *Fashionable Life* but a system of intolerable trouble, in pursuit of the reputation

Figure 4 Frontispiece of James Beresford, *The Miseries of Human Life* (London: William Miller, 1807).

of perfect ease?"[57] The characters then produced a series of riotous vignettes called "groans" that touched on the troubles with heating, drafts, coal dust, cold and damp, stifling interiors, and overcrowding. One Testy groan is about "[w]aking, stiff and frozen, from a long sleep in your chair, by the fire side: then crouching close and closed over the miserable embers, for want of courage to go up to bed; and so keeping in the *cold* to be *warm!*" Samuel Sensitive replies with the one about "the interval between breaking a pane of glass and the arrival of the glazier—N.B. The aspect of the apartment (your constant sitting room) E.N.E. and the wind setting in full from that quarter, at the crisis of the affliction—glazier a drunkard living seven miles off."[58]

The common link of unease and indoors in such snapshots testified that the domestic realm served as the battleground for mundane health. An economy of bodily care was virtually equivalent to modern domestic economy. For authorities like Sir John Sinclair, who was one of the leading health advisors at the turn of the nineteenth century, the purpose of studying the minutiae of domestic space such as the bedroom—"a place in which we pass the greater part of our lives"— was to establish the rules which the modern family should observe when deciding on the room's situation, size, ventilation, and heating. Drawing on Adair, Sinclair's *Code of Health and Longevity* brought into focus issues anywhere from the construction of the bed to the quality of feather fillings and bolsters, to the hazards of bed steps and charcoal warmers, bedclothes, bed-curtains, etc.[59] These concerns, in turn, gradually shaped practices aimed to provide "health" in one's interiors, and these, in their turn, began to change the overall identity of domestic life and the nature of "environmental labor" that this life was seen to depend on. To take a simple example, heating one's bed with warming pans or a stone was something of a standing order in well-to-do households in which even chambermaids had some "judgment on gaseous combustion."[60] The number of domestics, their work and presence in the household was predicated on the kind and intensity of the labor that was expected to keep up the environmental appearances of a respectable home. Households, in other words, were undergoing a socio-atmospheric overhaul in which the bedroom, kitchen, and drawing room assumed central importance as the sites where smells mixed with emotions and crowd with intimacy. The bedroom claimed attention for its rebreathed air, the kitchen for its olfactory gradients, and the drawing room for its partying, in which "more health [was] actually expended" than in the theater because "in our own, or our friend's drawing-room, we are so much at home, that any avoidance of draught, or precaution against exposure, is unthought-of."[61]

This was not quite true. An anxiety over the pathological properties of nocturnal air caused a widespread shutting of doors, windows, and wall crevices before retiring for the night. During epidemics, it was common to follow the time-honored policy of "aerial quaranteen" whereby residents "barricaded" themselves against the outdoor air by sealing away the contagious matter outside.[62]

The custom resulted in the much-debated morning smell and sickly humidity of the bedroom,[63] emphasized by the fact that bedchambers had lesser volumes and their occupants sealed them off from outside air by blinds and shutters, sash fastenings and pegs, and heavy window curtains and bed hangings. Some observers thought such caution came close to being a pathological condition of *aerophobia*—as "the dread of air, a kind of phrenzy"—a disorder that French physician François Boissier de Sauvages classified as a species of hydrophobia.[64] Struggling to overcome the smokiness from open fires, Benjamin Franklin confessed to have been a victim of the belief in the dangers of outside air, "aerophobia." Having realized, however, that no amount of moisture or cold would cause a disease, he converted. Embracing raw elements and rejecting fear from exposure was a symbol of enlightened mind: "I now look upon fresh air as a friend; I even sleep with an open window." But trite as such comments might have sounded to Graham or Wesley, the virtues of common air were less appreciated among the general populace, in Franklin's view at least. It is symptomatic that he thought this was still a case in the last decade of the eighteenth century: "You physicians have of late happily discovered that fresh and cool air does good to person in the small pox and other fivers. It is to be hoped that in another century or two we may all find out that it is not bad even for people in health."[65]

Sudden changes in air quality raised particular concern. Such risks had already been debated in relation to weather changes. Medical meteorologists argued that epidemic fevers followed after a moist spell succeeded a period of dry and sultry weather, or when in the winter, the southwesterlies changed to a sharp northeastern.[66] During such changes, a drop of temperature could cause internal regurgitation that slowed the blood, followed by inflammation and hemorrhages, eventually climaxing in fevers. The sufferer was one who was ill prepared to amortize drastic shifts. But increasing attention was given to dangers arising in the movement between the home and the outdoors. Writers on hygiene, longevity, and rules of everyday conduct warned of hazards from exposure to unexpected heats, cold blasts, damps, and drafts in and around the domestic realm.[67] Changes such as these undermined the equability of interiors at the time when improvements in heating and thermal insulation increased the gradients between inside and out: "[I]t is indeed probable that three-fourths of the disorders to which human constitution is liable in this climate, originate in, or are at least considerably influenced by, aerial transitions."[68]

Practitioners warned that thermal shocks compromised the comfort and health of the delicate who took thermal uniformity as a norm. "They commit a most dangerous error, who, in the winter nights, come of the close, hot rooms of public houses into a cold and chilling air, without cloaks or surtouts."[69] "Let him who would keep his health in this severe weather, direct his care to two articles," recommended Martha Bradley, a writer on cookery and the author of *The British Housewife* (1756), "first, the not exposing himself too much to the weather, and next the not locking himself entirely out of it. Cold and its consequences arise from too a careless Exposure to the Inclemencies of the Season; and nothing on the other hand is more unwholesome than too hot Rooms." Bradley's argument rested on the assumption that the air in all rooms heated by fireplaces was so heavy that it kept the body "in so unnatural a condition of heat that it is almost impossible to avoid catching cold in going out." Proper conduct required avoiding cold outdoors as much as overheated living quarters: "A Medium is best in all things."[70] Capillary damage due to cold known as chilblains epitomized this principle: They were caused by the exposure of feet and hands to the open fire after a prolonged period spent in the cold. Constipation, on the other hand, was a result of overheating by woolen clothing and long hours in bed.[71]

A medical concern for the physical movement through spaces with changing atmospheric properties was widespread. Physicians observed that health suffered when one moved between the countryside and the city, the continent and the sea, the north and the south. They also found that daily amplitudes compressed and rarefied humors: "[F]or if one Part of the Day causes a retention of the Effluvia, whilst another exhales them too much; so that whatever state they were found in the Evening, they are under very different Circumstances in the Morning: How far the Fibres and Juices then may be vitiated is not in the Power of Art to determine."[72] But domestic scales of exposure claimed even greater importance. Only the door separated a scorching drawing room and a freezing day outside. Indeed, the unequal distribution of heat and humidity were by far the most common among the complaints. To give an example, we may look at a letter written by the Irishwoman Emily Mary Fitzgerald, Lady Kildare, to her husband on December 9, 1762, which captured the complaint in following terms: "I have been here these two days and would you believe it? Starved with cold after coming from Castleton, which shows the coldest houses when constantly lived in will be warmer than the warmest when at any time uninhabited, as is the case here... You will say, what, was the print room cold? No, but the way from it from the apartment we are in at present perishingly so—those stairs running with wet, as is the passage above and most of the rooms to this back side of the house; which shows, my love, the necessity of having very, very often fires almost all over the house."[73]

Echoing such worries, a later writer stressed the absurdities "universally witnessed in our houses, of allowing the halls, staircases, and bedrooms to be of the temperature of the wintry blast, while our sitting rooms are kept at the summer heat." He was summarizing the consensual view that gradients of atmospheric hazards could span across space. English customs could induce foreign visitors to imagine "that so far from complaining of our *variable* climate, as we invariably do so, we were so enamoured of it, that we did all we could to imitate it within doors."[74]

But neither fires nor double doors conserved heat in the English house. For a country with no prolonged frosts, it was ironic that a critic of domestic heating was a Swede, Peter Kalm, who, while staying with a London merchant in 1748, observed that during February, the temperature of his host's drawing room never reached fifty degrees Fahrenheit. Swedes would never be able to "imagine that English cottages are colder (in winter) than Swedish [...] It is as cold indoors as outdoors."[75] Similarly, the architect John Wood found "that [many modest households] were unhealthy from the lowness and closeness of the rooms; from their facing mostly the north and west; and from the chambers being crowded into the roof; where having nothing to defend them from the weather but the rafters and bare roof without ceiling; they were stifling hot in the summer and freezing cold in the winter [...] the dormers leaky added greatly to the dampness, unhealthiness and decay of the cottage."[76] Anxieties of this nature mirrored the negative portrayals of Georgian domestic space, in which bad habits made it a source of risks that demanded action. Doctors advised against casual movements in and out of house and against the late-night use of carriages, in which the leather and lining imbibed dampness. Those prone to do so could be "instantaneously chilled by the sudden transition from suffocating heat to piercing cold."[77] Inversely, getting inside was an equal gamble: On frosty days, people were reminded to acclimatize in unheated antechambers. In all such advice, physicians took daily risks to the core of medical thinking about regimen, thereby expanding the remit of medical knowledge to the realm that had been rarely conceived as a space of risk or subject to rational control. Ordinary surroundings and private space were constituted as spaces of instant pathological potential that required surveillance and management.

Environmental Health As Domestic Labor

The prolific advice on indoor hazards was also a commentary on the material and spatial changes in the English home during the second half of the eighteenth century. During this period, middle-class houses typically used fireplaces and wick-lamp illumination. The smoke these produced sometimes combined with the kitchen fumes and mixed with body odors and vapors from drying clothes. Fireplace drafts and window construction allowed some circulation, but the air's admission was generally random, unintended, or a subject of taboo. But the medical awareness of indoor pollution also reflected the more general trends in sociability and privacy of the middling ranks that resulted in a gradual patchworking of spaces and multiplication of annexes that, quite literally, shrunk the average room's air volume. The changing function of rooms in country houses between 1720 and 1770—during the so-called shift from a "formal" to "social house" layout—involved an increase in the number of rooms in the state apartment.[78] In addition, the urban home was increasingly papered, carpeted, and curtained, features that lessened the airflow and increased the absorption of odors by books, pictures, collectibles, statuettes, and sundry trinkets. Such changes in "the physical environment of the middle class family [were] immense," yet what mattered as much was a change in the distribution of work that kept pace with the intensified house maintenance.[79]

One of the reasons for this change was middle-class affluence, which increased demand for servants. Throughout the century, the growing buying power of merchants, professionals, and gentry spurred a staggering new supply of objects for mass consumption. Some of these found their way into homes in the form of new furniture, decoration, and interior design. Urban dwellers in particular could afford more wooden floors, rugs and mats, drapes and shutters, upholstered furniture, hangings, paintings, wall candelabras, ornamental mantelpieces, and relief dados. More maintenance was required: more washing, sweeping, dusting, cleaning, polishing, tidying, and fresh water. Such work raised the levels and expectations of cleanliness and comfort by increasing "the frequency of washdays and the regularity with which floors and steps were scoured."[80] Clean comfort, in short, depended of the availability of domestics and the arrangement of their "environmental" labor: Fresh air and polished floors owed less to enlightened medical habits than to domestic labor and the control over quality of space, which was made possible by its availability in the marketplace.

Servant labor allowed affluent individuals to make their homes a stage of unfettered hygienic management: Performed with ritual care, airing and cleaning one's room, in their minds, symbolized propriety and discerning judgment. Especially important in this ritualization of cleanliness was the invisibility of servant labor: Enjoying fresh air during the party assumed that the air would be let in prior to the arrival of the guests. Or, expecting to enter into a heated bedchamber assumed it being heated prior to the arrival of the master. Adair recommended that a maid "stoved" the bedsheets every night to help evaporate the perspiration droplets and heat the sheets for the master.[81] For the master, the warmth resulting from the maid's activities made it look natural rather than manufactured. More generally, being relieved of the duty continually to recreate a "healthy" home by one's hands was the privilege of having other people's hands do so instead. Such a privilege was becoming commonplace during the eighteenth century. Dorothy Marshall showed that early modern England saw all members of a household sharing in housekeeping activities and the housewife was busy with cleaning, cooking, and provisioning on an everyday basis. By the early nineteenth century, however, traditional "housekeeping" had all but disappeared: "The noblewoman and the gentlewoman had delegated it to others, the laborer's wife, now often working in the fields, had little time and energy for housework and little enough to cook."[82] By this time, it was literally *easier* to create a healthy, aired, and clean home. The delegation of menial housework was at the heart of medical "naturalization" of orderliness, cleanliness, and the environmental comfort of eighteenth-century interiors.

Cleanliness equaled labor; filth equaled neglect. Being healthy and ventilated involved work and discipline. It did not represent a state of things, but an accomplishment. Medics argued that bad air resulted from idleness, and that only idleness could explain an "inveterate public insensibility to pollution, even after the way was shown how to enjoy the blessing of viatic purity." Naturalist and inventor Benjamin Thompson, Count Rumford, noted the difficulties in prevailing "on the public to accept the boon of improvement even in the matters which come home to every man's business and bosom." Hanway, too,

Figure 5 W.H. Pyne, "Groups of Women are Washing Clothes" (London: Pyne & Nattes, 1802).

sensed this difficulty when commenting on "how the use of these [principles of refreshing the air] is neglected in common life, especially among the indigent, as if they disdained to accept them at the hand of God." The poor in particular were slated for their unsanitary habits even as their own hygienic preferences could often be quite different. Some of them saw the municipal cleaning acts as the "acts of rigour and oppression, tending to sacrifice the privileges and enjoyments of the poor to the squeamish feelings and effeminacy of the rich."[83] When London's York and St. James streets were "paved in the NEW WAY, the mob were so displeased, that what was laid down in the day they pulled up in the *night*.[84] Counterintuitive as such actions might have appeared to an Enlightenment official, they were neither unusual nor limited to the town life. It was, for example, known that attempts to curb indoor smoke had to overcome the traditional rural belief that smoke hardened the house timber and even helped in the recovery of lung inflammation.[85] Contemporaries thus sensed a resistance to improvements that the poor thought were infringing upon their own (in)sensibilities and rights to feel comfortable on their own terms.

Such views contained a good deal of economic rationale: The primary concern of the rural population was to stay warm rather than be ventilated. Eighteenth-century newspapers abound with reports on charity given to the victims of cold: "We hear that the right Honourable the Earl of Tankerville has had a dinner dressed every day during this cold weather for the relief of the poor of Bradfield and all that are able to go, and those that are not it is sent to."[86] When severe frost cause the price of coal to increase from thirteen to thirty-four shillings, Sir Edward Gascoigne from York, hearing of the distress of local people, "sent from his coal pits three wagon loads one to the prisoners in the Castle, one to the prisoners in House-bridge Gaol and one to the Poor in Micklegate."[87] It was widely known that an insufficient supply of food rendered the body less capable of "bearing those changes of temperature, and currents of air, which would otherwise be agreeable. Protection from cold, therefore becomes an especially primary object of the poor."[88] On could argue that, given the economic condition, the general population could not afford to be medically enlightened. Acknowledging this financial origin of environmental "ignorance," Hanway acknowledged that the poor were not "naturally" dirty, but far too often cognizant of their misfortune: "[T]he young, the old, the virtuous, the profligate, the sick the healthy, the clean, the unclean, are huddled together, and inhaling a stagnated and putrid air, deplore their miserable situation."[89]

What, then, were the origins of the claims about the purported "inveterate public insensibility"? Why did physicians regard the general populace as insensitive, ignorant, and even subversive to rational medicine? In most cases, such claims did not only refer to poverty, squalor, stagnated air, and malodorous clothes. They also revealed something about the subjects who made them. They implied an elite "sensibility" to the minutiae of domestic life defined by a display of order and cleanliness. An orderly family life, with its concomitant separation of spheres, became the norm against which the middling ranks assessed the pathologies of aristocrats and proletarians alike. And yet it was not domesticity per se that served as a norm, but a domestic space as organized, controlled, and maintained. Only to the extent that it was sustained as a rational domicile did domestic space provide a norm against which to judge decay. "In Do-well's house all was regularity and neatness; you might set your watch by the time of their meals." A provident economy of domesticity thus ordered the social, geometric, and physical space, and coproduced bourgeois identity through leisure, sociability, hygiene, and interior design.[90] Furthermore, when the health writers criticized neglect and filth, they absolutized a wealth-specific availability of environmental labor and so permitted the enlightened minority to consider themselves in possession of a "natural sensibility" that defined standards of hygiene and regimen. To the extent to which it represented one of the properties of the nervous system, delicacy naturalized the "shock" felt by the delicate society when faced with conditions that they would refuse to tolerate in their own space. This reading gives a special meaning to Adair's recognition that the "affluence and luxurious indulgence expose us to distresses of a kind not suffered by the poor."[91]

It is somewhat ironic that weather as a medical entity remained marginal compared to the concern with a pathology of enclosed space. In this sense, the medical study of living space derived only partly from a Hippocratic interest in "airs, waters, and places." Its main impetus lay in the cultural critique of the changes in middle-class sociability. Its tool of analysis was the notion of a class-specific vulnerability to neglect. For this reason, the susceptibility to polluted airs reified—if not fetishized—the general anxiety over the risks in social sphere, because it assumed health in an image of a commodity under a threat. Paraphrasing William Coleman, the threat from the outside provided a "framework for articulating the primary demands imposed by the conditions of existence upon men and women who sought seriously to preserve their physical well being."[92] With this in mind, we may begin to see why the medico-moral concerns over the material circumstances of English urban living defined medical meteorology as a science of lived weather unfolding on domestic scales of exposure.

Chapter 3

Artificial Airs

Proper attention to Air and Cleanliness would tend more to preserve the health of mankind, than all the prescriptions of the faculty.
The Universal Family Physician and Surgeon[1]

By the last two decades of the eighteenth century, an attempt at air reform gripped the British scientific, medical, and political establishments. Medical analyses of smoky, damp, or rebreathed air shifted attention from air as a carrier to air as an immediate cause of disease. Physicians moved hazards from outdoors to indoors and, in addition to miasmas, began to discern the less spectacular if more tangible aerial poisoning in the everyday spaces of leisure, work, and sleep.[2] The atmospheric chemistry of indoor space inspired by the sociability of the elites and the gloom of the poor pronounced health a subject of mundane routine. Attention was therefore not always directed toward illnesses such as typhoid or smallpox, but on the "trifling disorders" such as headaches, fatigue, giddiness, and other discomforts linked to hygienic neglect or lifestyle excess. From the strictly medical point of view, casual indispositions and daily discomforts belonged to the analysis of the nonnaturals, such as drastic changes in air, diet, evacuation, rest, and exercise. By 1800, however, writing on the subject was also an analysis of a society engulfed in a maelstrom of unwellness. By this time, the analysis narrowed down from the general advice on the management of nonnaturals to the more specific interest in the meteorological life of the household. The signs of unwellness were increasingly more specific and increasingly indoor in quality. They included vertigo, dyspepsia, irritable bowels, bedroom cough, children's itch, rheumatism, nausea, diarrhea, and a loss of appetite. Combined, they threatened to cause a "transitory weakness of limbs," pallid skin, sallow countenance, or a flush of fever. Those spending time desk-bound developed disorders that stretched from chronic drowsiness to tightness in the forehead to torpidity of the senses. Doctors argued that if untreated, such feelings could turn into "serious maladies."[3] Many of such conditions were known as asthenic.

Physicians described asthenic symptoms as lethargy, decreased muscle tone, and infirmity of the nervous system. Cullen defined it as *"Languor au debilitatis corporis,"* or *"debilitias omnium artium, superstite actionum vitalium minutarum tenore, superstitibusque sensibus, sine dolore."* George Motherby's dictionary called it "extreme debility," and John Aitken grouped it with torpor, atonia, paresis, and tremor under the heading of palsy, and related to an "abolition or diminution of sensation and muscular action."[4] In his *Guide to Health* from 1795, Joseph Townsend identified it with the symptoms debility of the nervous and arterial systems, which he termed "hysterical diathesis."[5] Like unwellness, asthenia was manifested in a sluggish metabolism that marred mental focus and caused ennui.[6] It caused gory dreams; sufferers became fussy and whimsical, hectic and reclusive, gradually reaching the stage of a confirmed disease and becoming "through life, a burthen to their friends, the public and themselves." The disorder lived to become a model disease of the urban English, an early version of the fatigue syndrome that gained notoriety in the nineteenth-century "sofa cases."[7]

Asthenic conditions and chronic fatigue did not share a single cause. Diet, previous illness, inheritance, and constitutional proclivity played a role, but its progress depended mostly on two factors: inactive life and noxious surroundings. External conditions could aggravate its symptoms but their effects remained temporary: Its incidence could increase in damp, sheltered, low-lying places, especially if connected to "impure air, sedentary occupations, anxiety, and the irregular mode of living in a crowded city." In one of the early treatises entirely devoted to chronic fatigue, Thomas Withers, physician to the Edinburgh Infirmary, argued that the condition resulted from an overreliance on drugs and intrusive therapy at the expense of natural remedies and pure air. It started with indigestion and breathlessness, soon taking on a form of irritability, irregular pulse, and inflammation. Following up on James Gregory's critique of sheltered lifestyles, Withers spoke about an "unnatural delicacy" developing from the fear of cold and identified heat as one of the exciting causes.[8] He linked it with urban smoke and workplace pollution, identifying those employed in brewing, tanning, painting, dressing flax, and burning charcoal, as far more likely to suffer from its effects in consequence of being exposed to dust, steams, pungent smells, mephitic vapors, and metallic exhalations. Risk from asthenia increased in summer, when sunlight hit city walls and pavements to create a greenhouse glow and scorch the passersby, sparing neither the pampered and the sedentary, still less the poor and the under-sheltered.[9]

Prevention boiled down to environmental control and personal regimen. Doctors initially proposed a demographic density management: Withers, like later Adair, warned against "large companies and public assemblies in unventilated rooms," recommending smaller fires to lower the room temperature and control the dispersion of heat. Doctor John Fothergill urged patients to visit the country for inhalation of florid fumes. William Buchan advised living in high-ceiling apartments. Although exercise, rest, and cold bathing helped recovery, physicians assigned the highest value to exposure to fresh and equable airs. "[T]he frequent ventilation of rooms by opening doors and windows, is of great consequence to the restoration of health, and is a practice therefore to be strongly *inculcated*."[10] Withers chose to use the term "inculcated" with accuracy, because at his time, a ventilation system designed solely for the purpose of domestic comfort was above the means for a vast majority of middling ranks. Regardless, ventilation stirred an enormous public interest and soon lived to become an emblem of enlightened rationality and domestic ideals of comfort. The question is how.

Engineering Freshness

Ventilation has a vast history.[11] Conceived initially to combat fevers, its early eighteenth-century advocates described it as a solution in places plagued by overcrowding, humidity, stale air, and putrid exhalations. Solutions were proposed to correct the dynamic, chemical, and thermal features of indoor atmospheres. But clean heating and regular airing were necessities in a land cursed by the "extremely variable and uncertain" weather that compelled people of all ranks to "keep Fires to sit by near Eight Months in the Year."[12] Early machines such those proposed by J.T. Desaguliers, Stephen Hales, and Samuel Sutton were purpose-specific, intended to benefit politicians, convicts, soldiers, slaves, ship crews, and hospital patients. During the early 1730s, Desaguliers, for example, applied bellows to ventilate of the House of Commons. A decade later, Samuel Sutton found that the stench onboard Navy ships could be removed by continual fires, which rarefied the air in the rooms under the deck and caused a powerful draft through an unused chimney that freshened the room air.[13]

Stephen Hales developed a prison ventilation system in 1750 that he compared with the respiratory function of the organism: "[W]ere an animal to be formed the size of a large ship, we are well assured [...] that there would be an ample provision made to furnish the animal with a constant supply of fresh air, by means of large lungs [...] Can it therefore be an unreasonable proposal to furnish ships, gaols, hospitals etc. in the same manner with the wholesome breath of life in exchange for the noxious air of confined places."[14] Hales's lunglike machines became a medico-legal technology designed to purify hospitals, ships, and prisons in the way that French inventor Jean Noel Halle regarded his *ventilateur* a weapon against mephitic emergency as "an extinguisher" of poison by "a continuously circulating and pure body of air."[15] Attempts of a similar nature continued for the rest of the century, to include, among many others, William Cooke's 1752 plan for steam heating and ventilating greenhouses, and Richard Brocklesby's introduction in 1759 of ridge ventilation in decentralized military hospital units. While it is known that, in 1784, James Watt constructed a steam radiator to stimulate air convection in his study, most apparatuses operated in institutional spaces where they were expected to offset the emanations arising from tightly packed people.

If early ventilation was a social program, by the late eighteenth century it had become a commercial appurtenance. The baton of ambiental improvement was passed from philanthropists to entrepreneurial engineers. Ventilation could now perform several functions: It

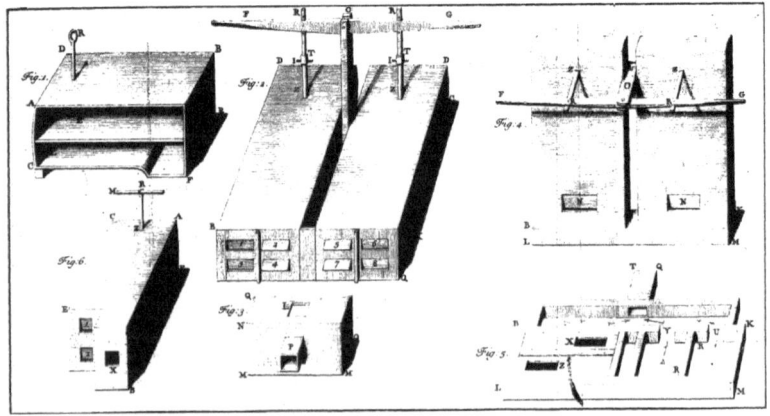

Figure 6   An illustration of the parts of Hale's ventilator in Stephen Hales, *A Description of Ventilators* (London: W. Innys, 1743).

enabled the house to breath with its surroundings, sheltered privacy and family comfort, and served against the "fixed air" buildup due to domestic fires and rebreathing. Influenced by the norms of comfort, health, and status, ventilation became associated with wealth and status at a juncture where the medical construction of space entailed a commodification of air by medical pneumatics, genteel sensibility, and engineering know-how. Engineers, vendors, buyers, fitters, architects, builders, medics, and physicists of domestic thermometry came to consider ventilation as an emblem of medical enlightenment rather than a newfangled innovation with a short expiration date. Given the scope of consumer culture by the late eighteenth century, such developments may appear logical enough. Yet thus far historians have mainly looked into ventilation in relation to the government's involvement in procuring purer air for those who lived or worked under special circumstances. However, to understand how ventilation became a symbol of modernity and progressive medical thinking, one ought to move from institutional settings—churches, prisons, theaters, hospitals, houses of Parliament, granaries—and examine its uses in domestic space.

Installation of early ventilation systems had to overcome technical, thermodynamic, and cultural obstacles. The majority of those considering using the technology raised issues about the environmental consequences that an artificial current of air would have on the indoors. For example, the volume and velocity of air that Hales's bellows brought down on the patients of St George's Hospital invited immediate complaint and resistance. Neither patients nor doctors knew whether the freshness of the incoming stream justified the lower temperatures and subjective feeling of chill. Many argued that ventilation made sense only if it guaranteed freshness without discomfort. Residential space was meant to be warm and cozy, not drafty and cold; permeability to external air was a builder's fault. For most people and for most of the eighteenth century, the avoidance of drafts, dampness, cold feet, and nocturnal miasmas far outweighed a philistine anxiety over the stuffiness of house airs. Jeremy Bentham captured this with sarcasm: "For the most ignorant feel the coldness of fresh air; and the learned only understand the necessity of it to health and life."[16] Such logic made sense especially with regard to the houses of the poor, in which ventilation seemed significantly less important (if not positively undesirable) than the warmth, insulation, and the elimination of moisture. Enid Gauldie has argued in her history of working-class housing that "the reformer's obsession with ventilation, natural in view of the medical profession's belief that most infectious disease was air-borne, took little account of the poor family's great need for warmth."[17] Neither should it be ignored that air ducts or ventilators required a major investment and a new approach to masonry.[18] However, even if all such difficulties could be overcome, and the combination of health and comfort not compromised, ventilation was suspect on grounds of its artificiality: "We do not understand what you mean by this new-fashioned air," the nurses told Hales, "we are contented with God Almighty's air."[19]

Withers had experienced this hostility when he confessed that any *planned* change of air had to be impressed onto people's minds. This was a shared opinion among most reformers who all noted a popular prejudice against medical argument, lamenting the lack of interest in "the right use and the management of the air."[20] Owsei Temkin captured this in more general terms when he argued that "as late as the eighteenth century the avoidance or removal of substances because of their potential harmful physiological actions has not yet become the leading concept in the idea of cleanliness."[21] But if popular indifference made sense among the needy, it seemed out of place among the prosperous. It would have been natural, hygienists contended, if the enlightened ranks led the way by introducing the system into their homes. Hanway wondered why "this friend of mankind [Hales] should have so few genuine disciples, and faithful followers, ready to furnish a liberal supply of this heavenly nourishment, by such ordinary means as reason suggest, or in some cases by extraordinary assistance?" In 1762, he thought it "amazing to consider how well the Principles of this Science are approved, how generally the Practice is known, and yet how ill it is attended to in many Instance. It seems rather to be the extravagant Fashion of the Times, to run counter it, and to poison ourselves."[22] Reformers like Hanway argued that people might have even wished to miss an opportunity to improve aerial quality of life even if they hoped, as did the architecture writer John Carter, that the remedy will "at length prevail over that unaccountable sloth, or obstinacy, which, where particular interests are not concerned, seems to posses the generality of mankind."[23] But the trope of popular "ignorance" often served to justify "improvement" in a way identical to that in which the trope of "closed air" served to justify "ventilation."

What the enlightened views ignored, however, was that the public distrust could never really be explained in terms of a lack of sensibility and mere ignorance. Even those who could afford to pursue Hales's or Hanway's ideals might have serious doubts about domestic use of machinery with an objective to purify air in jails or slave ships. In other words, domestic ventilation might have been a contradiction in terms. Why spend resources on airing a space used by a select group of wealthy individuals who would already have enjoyed the hygienic benefits of servant labor, fresh water, and plenty of room space? In highly maintained residences, ventilation could look redundant if not megalomaniacal. It could also appear misdirected, because the common wisdom about what constituted the air quality differed from expert opinion. Many residents knew that stuffy air was not the worst of indoor evils. In 1758, for example, an author of a health manual observed that the signs of bad air in a house were the "dampness and discolouring of plaster or wainscot, mouldiness of bread, wetness of sponge, melting of sugar, rusting of brass and iron, and rotting of furniture."[24] None of these resulted from the confined air, but from moisture, which was one of most salient problems in an average English home.[25] One could reasonably question the presumed benefits of letting even *more* of it in. Wasn't the purpose of the indoors a draftless and dry equilibrium of moderately heated air?

This is not to say that the medics shared a unanimous optimism about ventilation either. Generally speaking, many preferred minimal intervention combined with maximal avoidance. Fireplace drafts removed stagnant air while open fires burned stench and putrid fumes. Imperfect insulation and thin window glazing provided sufficient airflow without unnecessary technological complications. Architects recommended *perflation*—a method known in mining—in which "by letting Air in and out as they find proper, [miners] produce a kind of actual Circulation of it, and make it thicker and thinner, as they find best for their business."[26] In houses, perflation could be achieved by opening doors and windows, performed with care for the comfort of residents and adjusted to the house orientation, local weather, and topography. A "healthy" house was thus not judged solely in terms of its interiors, but also in relation with its location vis-à-vis external (in)salubrity. The healthiest residences were those built in the "Champagne" country and mildly rising grounds, but never high in the hills. They were to be placed on a gravely soil with "exposure prudently adapted to the nature of the climate," and with an eye on the prevailing winds necessary for effective perflation by means of drafts between windows. But drafts could be damaging if a house was built on damp, clay ground, in the vicinity of a marsh or a graveyard. While such topographic concerns influenced a choice of house site, they rarely posed problems after the house had been erected. If both the builder and the physician shared a concern about air quality, they also knew that it could be controlled by interior design based on the principle "pretty large but not too cold."[27] William Buchan, for example, advised that inactive persons "make choice of a large and well-aired place for study" that "would not only prevent the bad effects which attend confined air, but would cheer up their spirits, and have a most happy influence both on the body and mind."[28]

The Aeolus and the Sash

The healthy house symbolized progress and civilized society.[29] For midcentury medical scholars, prosperous clientele, and building authorities, it represented the achievements that separated the advanced from the brute nations. Sir William Chambers, Swedish born and perhaps the most traveled and "international" of English midcentury architects, argued that "in countries where men live in woods, in caves or miserable huts, exposed to the inclemency of seasons, and under continual apprehensions of heat, cold, tempests, rains, or snow, they are indolent, stupid, and abject; their faculties are benumbed, and all their views limited to the supplying their immediate wants; but in places where the inhabitants are provided with commodious dwellings, in which they may breath a temperate air, amidst the summer's heat and winter's cold, sleep when nature calls, at ease and in security; study unmolested, and taste the sweets of every social enjoyment, we find them active, inventive and enterprising."[30] It is worth noting, however, that Chambers explicitly linked his precepts with the maritime quality of national climate. In speaking about the distribution and size of rooms, doors, and windows, he pointed out the need to consider the economy of heating and so advised to make rooms "as close as possible." The windy and humid northern climates required that a room be planned to have fewer doors, none of which should be placed near the chimney, "as the opening them will disturb those who sit by the fire." Neither should doors be made on the sides of the bed, as this would expose beds to the draft let in through the crevices of the door. Layout consideration also applied to the placement of windows, the dimensions of which were limited by their glazing properties and permeability to cold. It is striking that in proscribing their height to no more than eight feet, Chambers was referring to the stately houses with ceilings as high as twenty feet. It is hardly imaginable that such volumes would require permanent ventilation.[31]

Even if it remains neglected in the histories of the environmental health, it can be argued that opening windows was the most common method of ventilation at any time in European history. It was a concrete manifestation of the beliefs about the relationship between "inside" and "outside." It was the litmus test of a society's (dis)interest in creating its domestic space in the image of outdoors. The door and the window marked the boundary where the vagaries of public weather met the private space of controlled warmth. Yet most of what is known comes from architectural theorists, for whom the climatological aspects of the house building dictated the window size and its position on the façade. Chambers's contemporary John Aheron proposed that in the temperate climate, the width of the windows compared to their "jaumbs"—the space between them—ought to follow the ratio of three to four, but should be wider apart (a ratio of three to six) in countries "more expos'd to violent Heat, or violent Cold." Aheron warned that this was only a rule of thumb, because the actual aspect of a building in relation to the sun and local winds "will always occasion a Variation in the Proportion of the Windows themselves."[32] Builders also knew that faulty window location was the main cause of smoky fireplaces.[33] In 1779, Robert Clavering wrote about the complex ways in which the outside wind and interior circulation prevented the evacuation of fireplace smoke, creating a "situation in life [than can be no] more uncomfortable and unhealthy." He insisted that both windows and doors required tight insulation and that no two of them should face each other.

Features defined both by affordability and legal requirements further complicated concerns like these. One of them was the window tax, introduced in 1696, which stipulated a flat-rate tax of two shillings per house and a variable payment depending on the number of windows over ten.[34] The baseline number was reduced to seven in 1766 amid complaints that the tax deprived inhabitants of both light and air. Bereft of daylight, the solution was a tallow candle, the acrid smoke of which became a staple of poorer households. Satirists linked the law to the Prince of Darkness warning the poor to "Open your Door, or go without it, and Air is free, you cannot doubt it: Nay, in your Windows there are Cracks, Where Air finds Passage—without Tax."[35] The tax became a medical issue with the rising concern over the living conditions of the poor. The meaner sort of houses with premises darkened in consequence of the window tax, "have been observed to exhibit a race of more pale and sickly inhabitants."[36] It could be associated even with fevers such as the one in Carlisle in 1778 when a local physician traced its source to a house tenanted by six poor families who had "blocked every avenue of light with which even wretchedness could dispense, to lessen the burden of the window-tax, and thus contaminated the air of their cells, to such a degree, as to produce the poison of fever among them."[37]

The medical significance of windows can also be traced in debates on workhouse sanitation, hospital construction, and the in-house treatment of the sick. In the last regard, physicians did not spare energy to advise on how to safeguard the fragile patient from the healthy but brisk currents of outside air. Allowing fresh air into the room of a patient suffering from contagious disease was always recommended for both subjective and hygienic reasons.[38] But often the concern was about the unwanted thermodynamic effects of the moisture and cold of the outdoor atmosphere. In his *Domestic Medicine*, Buchan spelled out detailed guidance on preventing window-related ills caused by the mistaken practice of cooling oneself by a window in an overheated room. In such situations, the current of air would be directed against only one part of the body, exposure to which the body could not properly adjust. In some cases, the practice yielded inflammatory fevers and severe tonsillitis. Contrary to a majority of doctors, Buchan also preferred that overnight windows remained shut even in the hottest season. Whether for fear of nocturnal moisture or malodorous city breezes, contemporaries regarded this a pernicious custom to be avoided even in the harshest weather. Doctor James Lind, among the most respected practitioners and an authority on naval diseases, insisted that no weather could ever provide an excuse for "shutting all the inlets of air," especially in prisons and hospital wards. But he noted different attitudes among patients suffering from different ailments. In smallpox and fever hospital wards, where windows were kept open all day and night, even during frosty nights, the patients rarely complained as long as they enjoyed warm bedding. On visiting recovery wards, however, Lind noted that the patients there could not be prevailed upon to accept such a practice: "[N]o sooner are the doors and windows ordered to be open for the benefit of a little fresh air, than the men are excessively chilled, and complain of the cold, even in moderate weather, being intolerable."[39] In a case of near desperation, Mancunian John Ferriar resorted to the radical measure of pane removals, which he felt was necessary to counter the practice of "putting up fixed casements." He argued that at least one portion, if not the whole window, should be mobile: "From the want of such a regulation, I have been often obliged to order several panes to be taken out of the window of a fever-room, to obtain a tolerable degree of ventilation."[40]

Adair thought that private residents should adopt a seasonal discipline in controlling the ingress of external air. During the warm and close summer weather, the advice was to leave the chamber door open for a few nights, and afterward only part of the sash. However, the shutter should intercept drafts. As residents got used to this level of exposure, the idea was to open the shutter completely but drop the curtain and reduce draft. During the winter months, Adair recommended leaving a window open in the room adjoining the one in use, but keeping an eye on the diffusion of air by the intradepartment doors.[41] In workhouses, on the other hand, cross-ventilation by opposite windows was recommended by the Quaker John Scott as early as 1770s, followed by Edmund Gillingwater's suggestion that such windows "should have a casement which should always be left open, whenever the weather permits, to promote as much as possible the free circulation of air through every part of the house, so essentially necessary for the health and comfortable accommodation of the paupers." This, however, remained difficult to achieve before the mid-nineteenth century, when cross-ventilation became a part of pavilion hospital design.[42]

Meanwhile, the general population vacillated on the merits of sash construction. Benefits were thought to depend on the season and heating sources; reports from before 1800 reveal reservations about the random admission of external air and the disturbance of indoor calm. Physicians warned the public to curb indoor air currents for two reasons: They were unpleasant when felt and dangerous when insensible. With the exception of those like Lind, many advised against the foolhardy exposure to such airflows and placed patients in sheltered areas of the ward. Any changes to thermal equilibrium required planning and caution. Children with ague were put in a "dry and warm bed, which should be in a large room, free from the draught of air. When circumstances and the season permit, the door or window might be kept open when the patient is free from the fever; this however, ought to be in such a manner that the patient should not be exposed to sudden gusts of air."[43] Lung disease was considered to increase in morbidity among those sleeping "near an open window, and by other partial exposures to currents of air, which seem to be more injurious than a general exposure to cold."[44] The febrile and those in tropical climates were liable to catch a cold due to heavy perspiration. A random breeze could bring on a fever, facial cramps, stiff neck, and joint pain. "When a lady, who has been heated by dancing, either sits near to a window, through which penetrates a cold and partial draught of air, or [...] when a reaper reeking with sweat, either drinks cold water, or lies down to sleep upon the humid grass; the injury is perceived before the cold has alternated with heat."[45] "How then must it be with a patient, who in a little hut of an hospital, is placed at a door way, or in a current of wind, or raised on a platform to the level of an open window, to prevent suffocation from heat, if a critical sweat should break out? The sweat is suddenly stopped; and if death do not ensue, the disease is lengthened out into months."[46] Concern of this kind was only confirmed with cases such as that of Beddoes's patient who "on merely approaching a window in severe weather set immediately to sneeze, and find all the symptoms of catarrh supervene in their order."[47]

But by Beddoes's time, the olfactory legacy of overcrowding had become a pariah among practitioners like Hanway, Adair, Buchan, and Willich. Generally, physicians, architects, and philanthropist concurred in thinking that the corruption of cameral airs amounted to a crime in a land blessed by the invention of such simplicity as the sash-window. The sash window—the most common form of window in vernacular, urban, and elite architecture during the eighteenth century—was imported from Europe to England during the 1670s. It usually consisted of three vertical frames of four squares each proportioned to the shape of the window, two parts of which were suspended on pulleys and could be moved with ease.[48] The mechanism could keep the window open in any required position without external support from catches or stoppers. The advantages of this "continuous" system led to what historians Hentie Louw and Robert Crayford describe as an "incremental ventilation control" that allowed for a crack to be opened at top or bottom of the window to admit the desired amount of air without danger of a sudden blast. The window would also stay at the required height regardless of wind and without fear of guillotining a resident. In this regard, Adair advised "wealthy persons to inhabit and sleep in lofty large apartments and have all their windows to slide down from the top; and unless the weather be very boisterous, to open the upper part of one to renew the air frequently."[49]

Airflow control preoccupied both users and constructors. The latter tried to direct it by an upstand placed at point where the bottom rail joined the sill to permit "trickle ventilation along the meeting rails when the lower sash is slightly open."[50] Construction elements thus enabled the sash to prevent the buildup of stale air, a feature well known to landlords and servants alike, the latter being instructed that "[w]hen you bar the Window-shuts of your lady's Bed-chamber at Nights, leave open the sashes, to let in the fresh Air, and sweeten the Room against Morning."[51] Hanway had already dwelt on the topic. He and Hales both thought it "amusing how the difference of an inch or less of the upper Sash, will change the quality of the Air without the least injury to any one present." However, Hales observed that the practice would work in a sale or coffee room, but was insufficient for a workhouse, school, or hospital, which required maximum air exchange. Indeed, outside the home, beneficiaries of ventilating machines include merchants whom Hanway described as "extremely distressed, indeed half suffocated, at Garraway's coffee house, at a Sale which has lasted 3 or 4 Hours, merely because they were afraid of taking this Method, though they might with great Propriety have sat with their Hats on."[52] Hanway also expressed concern about the condition of sedentary professions, generally expressed in relation to ailments of the deskbound.[53] Such considerations only highlighted fears from sudden exposures to air change, which continued as the main hindrance to popular acceptance of ventilation. In hospitals, both staff and patients protested against the artificial, downward gusts from air ducts. Hanway showed good sense publicizing the more "natural" sash opening as a safeguard against any injury caused by mechanical ventilation. Even then, he was sorry to report, they remained "extremely repugnant to set windows open [choosing rather] to droop and die with heat, rather than suffer any sensation of cold [...] It is possible, so to taint the blood, that the very purity of the air shall become morbific, till the patient is gradually habituated to it."[54]

The question is whether this "habituation" could explain the apparent resistance to ventilation on the part of the poor and the ill. I have argued that an "objective" state of being "used" to bad airs or ignorant about its proper handling must always be seen as a result of economic circumstances in which one had to become habituated to what one could not change. It is no wonder then that the first uses of ventilation came from progressive professionals and wealthier households. The example I use is Thomas Tidd, a servant to the Princess of Wales. Clearly Tidd was familiar with the state-of-the art domestic appliances and privy to the meteorological conditions of princely interiors. Early in the 1750s, he set out to construct a gadget designed to clean the air of rooms "in a manner much more commodious" than before, designed for the benefit of "the Public in general, and to the nobility and gentry in particular." Constructed as a freely circulating fan fitted into a metal frame, it allowed the flow of air as a result of very small pressure differences between the inside and the outside of the house. In *Considerations on the Use and Properties of the Aeolus a New Invented Portable Machine for Exchanging and Refreshing the Air of Rooms*, Tidd announced that he had obtained an exclusive right to sell the machine for two guineas (for a plain-metal version) and three guineas (for a gilded and ornamented one). The Aelous thus catered

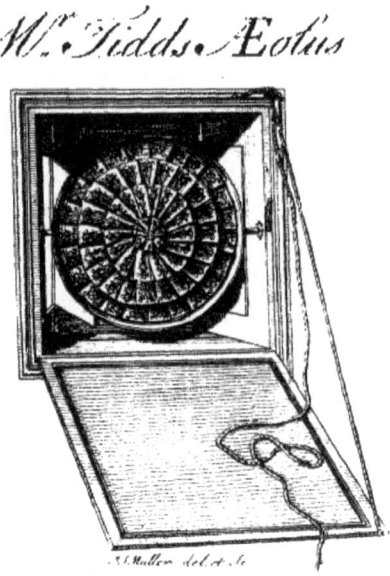

Figure 7   An image of the Aeolus in Thomas Tidd, *Considerations on the Use and Properties of the Aeolus* (1755).

to the ecologically inclined metropolitan society who needed to be reassured that their money was not spent in vain. Tidd's list of subscribers included even Hales himself, in addition to Prince of Wales, Prince Edward, and 110 gentleman, physicians, and lawyers.[55]

The machine, named after the ancient god of wind, was to be fitted in a square of a sash window, taking special care to minimize its protrusion from the exterior of the window. Tidd took particular care to make the Aeolus noiseless, automatic, and maintenance-free. Its biggest selling point was that in its mode of operation, it automatically stopped when it "sensed" an open door or a window. Tidd thought this feature constituted a major leap forward in the low-budget regulation of closed airs, as the machine "does not let it in with a gust, stream, or draught, as happened with a door or window is left open, but diffuses and spreads it in an equable manner over the whole room, so that there is not the least hazard, or even possibility, of catching cold, to the most delicate."[56] Tidd hoped that his "salutiferous" Aeolus could prevent such ills if it were allowed "in sick chambers, but also in all the rooms where the people assemble both public and private." Tidd launched a sales talk that portrayed the Aeolus's looks as meshing with the refinement of its quiet, automatic fan, the discreetness of which appealed to the delicate, learned, and wealthy. He put a premium on its ability to produce "the air to that exact degree of temperature that is *most desirable*" to the house inhabitants. Tidd's juxtaposition of use with style might have looked odd, but his invention soon proved to have no insignificant success.[57]

Improving on Nature's Climates

Designers' impatience with "vulgar errors" underscores the reality that household ventilation remained a minority interest long after its institutional uses found support among prison and hospital reformers. The lag may mean that general opinion viewed domestic air quality as too idiosyncratic to be tackled by medically informed gadgets like the Aeolus. There is little evidence that the population saw the matter as anything more than a subject of personal initiative and common sense. Even among progressive minds who could afford it, privacy remained sacred and the attempt to monitor it in overtly rational terms was seen as an encroachment on individual freedom. It required neither medicine nor the nose to argue that domestic atmospheres continued to be far purer than those found in what Christine Stevenson termed "closed institutions" such as hospitals or jails. In some cases, the air in private homes was positively salutogenic: Gilbert Blane, another of Cullen's great students and a statistician of naval health care, wrote in 1799 that he had witnessed more success in the treatment of compound fractures and violent injuries in private airspace than in the promiscuously rebreathed air of hospital wards.[58] Clearly, house air was different from institutional air. The former was about convenience rather than infection. Indeed, how could one prove a causal connection between smallpox and an overheated room?

More important than purely medical concerns in making ventilation a "natural" need was the triangulation of comfort, convenience, and social life.[59] The idea of comprehensive control of indoor atmospheres—including heating, cooling, ventilating, and maintaining constant temperature—drew in many upon the techniques used in the continental and British hothouse technologies. Artificial climates achieved in the so-called "forcing-houses" allowed the horticulturist to create summer conditions in any season and improve the quality of exotic fruits, for which there was an increasing culinary demand. The problems under discussion were worthy of the notice of Boerhaave, Linnaeus, and Adanson, who became involved in trials to correlate the roof slope and latitude and the control over air motion during the final ripening of fruit. Among the British practitioners contributing to the problematic was Philip Miller, the most distinguished British gardener of the eighteenth century and the superintendant of the Chelsea Physic Garden of the Society of Apothecaries. In relation to his extensive expertise in roofing and greenhouse ventilation, an early nineteenth-century historian remarked, "Had the demand among the opulent for early and high-flavored fruits been as great as it is now, it is probable that Miller would have worked more on maintaining an artificial climate during the whole year." Botanical hothouses, originally intended for experimentation, gradually became articles of trade and status. The know-how used in achieving their environmental control was one of the sources of domestic ventilation later in the century.[60]

One of the early applications in which the greenhouse notion of a permanent, rather than a discreet control of indoor air was that of James Anderson. With a residence in Edinburgh, where he had attended William Cullen's lectures in pneumatic chemistry, Anderson was one of the leading political economists of his time, a collaborator with Jeremy Bentham, and an accomplished author on agriculture and engineering. An avid improver, Anderson's seminal ideas about aerial management appeared in his *Practical Treatise on Chimneys* (1776), in which he endorsed the traditional use of fireplaces as the simplest method of draft-based ventilation. He explained that in a room heated by an open fire, any attempt to feed the fire with an aided airflow would stop "the perpetual ventilation" through the crannies and crevices. But while this would allow for the ventilation of the lower part of the room, the warm and wasted air would accumulate near the ceiling. One solution to this would be installing vent holes in the upper part of the sash, which would, as a side effect, encourage smokiness in the room. The other, preferred option was an advanced system of circulation involving an opening in the ceiling of the room through which the warm air would be removed, by means of the wall pipes, into any other room of the apartment, or released outside the house. In this semiclosed network of pipes and reservoirs, as soon as the warm air left the room through the ceiling pipes, the cool external air would enter in through another set of openings, mixing with the ascending warmer air until it would imperceptibly "reach the candles and the company in the room [...] without the inconveniences to which the company are subjected by the usual way of admitting fresh air."[61] In Anderson's view, this would allow for a perpetual control of the house environment, "without endangering our health on the one hand, or, on the other [...] being exposed to such exceedingly unequal degrees of heat and cold,"[62] as was unavoidable with common methods of perflation by doors and windows. The important consequence of this proposal was that the healthiness and comfort of domestic space could be thought of as depending on technology rather than the natural "permeability" of a house based on the manual labor involved in scheduled perflation. By these means, "a climate might be produced more perfect than any in nature."[63]

Automatism and user friendliness were as important as health benefits. The roof-mount ventilation pump that the Marquis J.B.M.F. de Chabannes installed on top of his town house at 1 Russell Square in the early 1810s was part of a grander vision to construct "new houses, in which all calculations and particulars have procured a very great economy and many delights." Chabannes became involved in all branches of heating and airing technologies, from reducing smoke in hothouses to installing a system of ventilation in the House of Commons and elsewhere.[64] His apparatus in London was a part of the vision to erect a building in which the air and heat, running through metal tubes, would open and close the windows or unlock the doors by pneumatic mechanisms. In the Russell Square house, a network of pipes connected each room to a recipient (collector), which drew in outside air and then circulated it around the house. This enabled residents to avoid the direct impact of outside air and created a more tolerable environment for lung patients: "[I]n [the house] any artificial atmosphere may be made and will then ascend into the apartment of the invalid." In proposing such an elaborate and expensive setup, Chabannes emphasized the convenience of a system that operated without "exposing persons to the perfidious currents of the air of chimneys, or to the variations in the exterior temperature, in fine without exposing them to a hundred accidental causes of a disagreeable influence to which the common mode of Ventilation is subject to."[65] It is of importance that Chabannes did not dwell on the removal of stagnated air. He rather planned for total control of air quality that would eliminate all "accidents" of both an external and internal nature, including odors, drafts, and smoke as well as outdoor humidity, cold, and heat. Fighting foul air and climate amounted to developing a intelligent, breathing house "as the only means of counteracting the effect of [the] rapid atmospheric changes, in the interior of our dwelling houses," to be achieved by a "judicious application of artificial heat, on one hand, and by an adequate degree of ventilation on the other."[66]

Many systems of similar construction were favorably received. Around 1790, William Strutt, a philanthropist and active social reformer from Derby, placed on the roof of his house a tube with a funnel mouth that faced and "collected" the incoming wind. The tube led to an underground reservoir a hundred meters away from the house. The difference in atmospheric pressure between the funnel and the container then forced the collected air through an underground pipeline back into the house, cooling it in the summer and heating it in the winter. Simpler and more commercial contrivances also became available. One was the "Air Machine" designed by a William White from Garlic Hill, London. Resembling a piece of furniture, it was efficient enough to cool the air in the sultriest climate, "a luxury of all others the most to be desired."[67] The apparatus mimicked the fine-tuned operation of sash windows by allowing the quantity of air and the degree of cold to be "increased or diminished at pleasure, and regulated so as to answer a great variety of purposes." The apparatus was advertised for use in hot *and* cold climates, on ships, in hospitals, and in all public buildings. If needed, the inflowing air could be perfumed to curb mosquitoes, presenting to the "people inhabiting our East and West Indies, a source of health and comfort, of all others most desirable, and to which they have hitherto been strangers."[68] Press releases were complimentary and the machine received kudos even from the directors of the Human Society for its applicability in resuscitation. It was praised by the Bath Society for the Encouragement of Arts[69] and by Admiral Dalrymple for its performance onboard *HMS Sandwich*. It was installed in Parliament in Dublin, London's Custom House, New Theatre in Drury Lane, East India House, and the Stock Exchange.[70] Comprehensive undertakings like these influenced the ideas about interior design, private comfort, and domestic engineering, and are best seen in the context of the changing architectural theory during the early nineteenth century. George Teyssot has argued that the architect's intention in this period was to transform the "habits" of future users rather than follow their wishes. If eighteenth-century architecture "spoke" and acted "through its form upon perception," nineteenth-century architecture was "moralizing" and "acting to reform."[71]

A Pedagogy of Disgust

Reformists argued that any doubts about man-made atmospheres only reflected ignorance about how long-term exposure affected health and transformed casual complaints into chronic maladies. The commercial success of domestic ventilation depended on the justification of its medical necessity in spaces where contagion, dirt, and overcrowding did not present imminent risk. Central to this success were the claims linking fresh air to social progress. Among the leading champions of residential ventilation was Anthony Meyler, medical practitioner, lecturer, constructor, and writer of a landmark apology for medical uses of artificial air. Meyler's education was from the University of Edinburgh, where in 1803 he defended a dissertation on mental depression. He then embarked on an engineering career, during which he installed at least a dozen ventilating systems nationwide.[72] By the late 1810s, his systems were operational—and praised—at the Liverpool Workhouse, Liverpool Dispensary, Liverpool's Jordan Street and Sunday School, London Chapel House, Stoneyhurst College, Covent Garden and Haymarket Theatres, Tobacco Warehouse at the Port, and Dublin Royal Barracks.[73] Meyler canvassed his vision in lectures before the Dublin Society in 1818, where he defended his products as the tools of health that he revalorized to meet his ventilating ideology.

If Meyler read Beresford's critique of domestic nuisances, he must have sympathized with the tribulations of urban dilapidation, clutter, and squalor. Meyler, like Beresford, developed his philosophy of viatic purity as an answer to the environmental side effects of daily life, which, unlike Beddoes, he refused to condemn as such even as he acknowledged their health hazards. Meyler argued that the quality of life owed more to curbing privation than extending possession. Happiness depended on the removal of "small but perpetual sources of minor inconveniences." It depended on trifling pleasures. Health was neither caused nor preserved by the "occasional operation of extraordinary and powerful agents." It was sustained by the constant but imperceptible influence of "background" causes of disease such as air, because "surely no agents can be more powerful in contributing to the health, than the purity of the air which we respire, and the well regulated temperature of the medium in which we live."[74] In fact, all environmental factors, regardless of their intensity, could lower the "general standard of health" and render the body more vulnerable. Meyler believed that lived space presented its dwellers with a myriad of risks whose cumulative power overshadowed the effects of acute diseases.

The medical judgment about the quality of atmospheric life was never solely determined by measurable entities. Taking the "objective" state of environment as a given would hide a moral framework that precedes the aesthetics of pollution or discomfort. I suggest that the aesthetics of stale air and thermal comfort, purity and filth, and health and disease contain an unsaid assumption about how objects and spaces become unwanted and intolerable. Following the insight from the preceding chapter, it is important to emphasize that Meyler constructs areal threats in terms of neglect: "When [an impartial observer] finds that impure air is the most fertile source of many maladies, improperly attributed to other causes, and that the means of removing this evil have not been resorted to, even in private dwellings, he will cease to declaim against the inhumanity of the age; but the *indifference* thus manifested with respect to health, will afford a *new and fertile subject for surprise* and reprehension. He will see how slowly useful improvements advance into general adoption, how much man is the creature of habit, and how custom blunts our perception of an evil and causes us to bear its inflictions without seeking to remove it, disposing us rather to regard it as one of those unavoidable visitations of providence, which man may deplore, but cannot prevent."[75]

Denouncing habit and custom, Meyler highlighted civilized sensibilities. But he also remained unconvinced that the latter had taken hold of those who should provide examples. He witnessed that the diseases found in the hovels of the indigent and the apartments of the poor also harassed the residences of the "luxurious and opulent." Irrespective of size, purpose, and design, pestilential air scourged "our prisons, our hospitals, our factories, our schools [...] our churches, our theatres, our place of business and of amusement, our courts of law, our houses of parliament, the palace of the monarch, as well as the residence of the artisan." Standards of cleanliness might have generally improved, but the state of dwellings proved that "many removable causes of disease are yet permitted to exist, and while all proclaim danger, nothing has been done to remove it."[76] Meyler had in this way presented the universal putrefaction of British indoors as an inexcusable moral fault and connected it with a technological fix. Using medical idiom and moral rhetoric for profit, Meyler's etiology established the technological solution of the problem that he constructed as preventable! In his view, a fatalism fostered by custom overlooked action mandated by available technology.

Removability entailed responsibility. Meyler first pointed out a glaring absence of government involvement especially where the cause of disease could not easily be identified. The people needed government to guard them "against those causes of disease which operate unseen and which therefore are not manifested to the watchful instinct of self-preservation." Government could either make such causes visible through education, or curb them through benevolent coercion. But Meyler also feared that even these would prove short as long as there was not enough evidence to sway public opinion in favor of prevention and top-down regulation. The purported medical risks linked to foul atmospheres were, in the minds of the majority, "incompatible with the fact that people live with apparent impunity in ill ventilated places, and if impure air be so destructive to life, how can they continue to enjoy health?"[77] Adaptation to local conditions and thus becoming immune "to the action of noxious agents" had been known among the northern nations' ability to endure cramped existence and levels of carbon dioxide far surpassing those found in the worst card rooms Adair saw in Bath. Such conditions affected unaccustomed visitors, but indigenous populations, like the English manufacturing classes, seemed to enjoy good health. If health was absence of disease, such an adaptation made ventilation arguments pointless.

Meyler initially responded that despite adaptation, no one should doubt the benefits of temperance, warmth, and wholesome diet. Meyler's full answer, however, had to wait until he could provide a working definition of health. For example, he doubted that the influence of bad air was necessarily dangerous. "There were many degrees on the scale of health, and there is a long interval between the degree which marks the full and perfect enjoyment of this first of blessings and the absolute infliction of disease. The great and important question of health must also be viewed with reference to our capacity of accommodating our constitutions, to varieties in temperature, diet, and modes of life."[78] Health and disease were about adaptation to external conditions and not about absolute states of the body. Health and disease were contextual entities that made sense only if the body was analyzed in relation to modes of life, its surroundings, and exposure to local stressors. Health and disease were the manifestations of displacement and maladaptation. Meyler's view signified a move away from the management of the body toward the management of the body in its milieu. Health and disease were only partly about the immediate causes of disease and more about an ability to cope with the external hazard. Because morbidity mirrored the lived space, medicine was to go beyond the care of the body and become a surveillance of external risk. Medicine required nutritionists, engineers, topographers, meteorologists, and philanthropists to work through the thickets of external influences on human health.

Like his contemporaries, Meyler referred to these conditions as predisposing causes, of which impure air was his main concern. Despite being careful not to assume that every whiff of stench damaged the lungs or caused diarrhea, Meyler produced a gory list of aerial diseases.

Bad air caused the stomach to reject food, enfeebled the general habitus, increased irritability, triggered fevers, and aggravated preexisting illnesses. Badly aired nurseries predisposed children to developing crooked shoulders, swollen bellies, large heads, pallid cheeks, sunken eyes, loss of flesh, and the lack of "infantile vivacity."[79] Importantly, however, Meyler argued that guarding against these conditions could not be done merely by the administration of common outdoor air. Especially in the cities, the idea of outdoors was fraught with difficulties. Furthermore, city people generally suspended environmental expectations they would otherwise maintain in the country: The city was *meant* to be polluted and urban dwellers were *meant* to adjust to it. This was not Meyler's belief. For him, healthy space had to be based on an absolute baseline that could only be approximated by outdoor air. This baseline assumed the air that was at once oxygenated, dust-free, warm, and draftless. Not found in nature, it had to be made. Anticipating objections, Meyler accompanied his argument by a series of thought experiments that amounted to a form of pedagogy that could be used to buttress the medical importance of mechanical ventilation.

The strategy involved imaginative excursions into the sticky air of the average bedroom, during which the reader could experience its invisible contents and learn to appreciate the nocturnal chemistry of rebreathing. The emphasis was always on the fact that such contents would remain unfelt as long as one stayed in the bedroom. To get the point across, one was asked to leave the bedroom immediately after getting up and the return to it in a minute. The routine can then be used an elementary building block of sense reaction and used as a template for an identification of larger-scale problems. An imaginary routine—"get up, go out, and get back"—proved how a seemingly inoffensive space could spawn stifling airs in matters of hours.[80] The method was effective because it invoked a sensation common enough to be ignored and real enough to be objectionable. Its purpose was to show that good air was not a given even in a solitary place like the bedroom, and that it needed to be perceived through a comparative experience based on the instinctive repulsion to body odor, stuffiness, and humidity. A comparative pedagogy of disgust became an experimental proof for the necessity of ventilation.

This was a known rhetoric. To imprint a medical argument about air quality onto the public mind usually involved a fictional individual being chaperoned through daily routines and asked to reevaluate the nuisances encountered along the way. The person's comparative experiences, real or fictional, could then be used to assert the reliability of personal feelings and empower the individual to take aerial health into his or her own hands. Tidd, for example, agued that "even the most healthful, who have frequented places where numbers of persons are assembled together, must have been sensible, from the headaches, and lassitude they have experienced in themselves, from the fainting and disorders they have observed in others, that confined air is extremely prejudicial to their constitutions." Adair also used it to explain that "the most neat and delicate person after having slept in a small bed-chamber does not, when he quits it, discover any offensive smell, but when he returns in a minute or two, and before fresh air is admitted, he will quickly discern a difference."[81] Hygienists knew that the air in a closely shut room could only receive so much of the perspired matter, and that after this, any additional matter would remain trapped in the body and cause uneasiness, slight irritation, and restlessness, "which is difficult to describe and few that feel it know the cause of it." Those using too many bedcovers would often wake up fidgety and restless, only confirming the need for fresh air even during the night. The author of "The Art of Procuring Pleasant Dreams" (1797) wrote, "To become sensible of this, by an experiment, let a person keep his position in the bed, but throw of the bed-clothes, and suffer fresh air to approach the part uncovered of his body, we will then feel that part suddenly refreshed for the air will immediately relieve the skin, by receiving, licking up, and carrying the load of perspired matter."[82] Learning by doing was essential, because, as Meyler put it, "If we come from the open air into a crowded room or an unventilated bedchamber, we will at once perceive the impure state of the air." It was crucial what happened next. If the person left the room or forgot about her feelings, she would ignore a physiological evidence of great value. Meyler thought that our sense of unease deserved medical attention.

The pedagogic uses of personal discomfort, it might be argued, constituted a historical crucible of environmental medicine in which innovators like Tidd, Chabannes, and Meyler laid down the epistemological ground rules that legitimized a technological vision of hygienic comfort. They found such rules in a democratic appeal to common sense, arguing that a neglect of personal discomfort—a "subclinical" headache, fatigue, or a sense of oppression—was an error perpetrating the ambiental bigotry behind predisposing causes of disease. Bedroom air was one of pillars of early "environmental" medicine. Most important, however, such experiences demonstrated a "naturally" felt need for intervention: "[W]e disregard the still more decisive evidence of the headache, languor, oppression, fainting, and other inconveniences, which are experienced in theatres, ball rooms, churches, and other crowded places, which are caused by the impure air we are compelled to breath, *in consequence of their defective ventilation.*"[83] Innocuous as they may sound to modern ears, such claims connected disease with a lack of preventive technologies. The medical problem was not in the actual oppressiveness of the living space caused by rebreathing, overcrowding, smoking, or dancing in overheated rooms. The problem was how to incite the public to *connect* the cause of disease with the absence of means to curb the side effects generated by such actions. Foul air ceased to emanate from the unwashed multitudes; it was *created* by bad ventilation. This, in reality, was just one of the means an eighteenth-century innovator could employ to advance his commercial interests. John Styles has pointed out that product innovation was not just a matter of bringing "unfamiliar products to market, but involved the formulation and reformulation of product definitions and identities in such a way that new products were rendered comprehensible and attractive to consumers." In the present case, the product was constructed as necessary once its absence was seen as the source of the problem.[84]

In such sales talk, a good portion of domestic discomforts and aerial diseases became *removables.*[85] But as removability became a leading trope in public health activism, it also naturalized the need for technical prosthetics. Meyler was of the opinion that modern generations ought to resign themselves to the facts of life they could not control. But he also thought that they could make such facts more tolerable. "If fashion continues to exert her imperative sway in crowding our apartments," he noted, "humanity, as well indeed as self preservation, should prompt us to make a necessary provision, that they shall not become unwholesome from an almost total want of that indispensable pabulum of existence."[86] Solution justified complacency. For Meyler, this meant that the preservation of public health must move from theory to action, from declarative rectitude to social reality, and from moral medicine to pragmatic engineering. When he wrote that engineers showed benevolence more than victims showed concern for their own condition, Meyler effectively placed the "bottom line" at the pedestal of social reform. In medical discourse on foul air, the comparative pedagogy of disgust demonstrated that the solution to the ills of modern life lay within the boundaries of human ingenuity. Such discourse also assumed that intolerance to conditions increased in proportion to a failure to act. Intolerance, in other words, was never caused by the nature of things alone—no matter how horrendous they might have appeared—but by a recognition that such things stood for social inaction. The possibility of prevention made discomfort still more intolerable. Writing from a Victorian vantage point, Walter Bernan, in his exhaustive two-volume history of ventilation, made a rare effort to capture a technological determinism underlying environmental medicine when he wrote, "[I]n most cases the physician is prevented from recommending a patient to be placed in air of a certain quality, from a supposed difficulty of producing it; or the friends of the invalid are indolent, and disposed, when anything beyond mere household means—such as warming pan, or hot blankets, or hot water—are required, to consider every thing not comprehended in the ordinary routine of practice with dislike [...] In other cases the relatives are poor, and unable to meet the expenses; or they are parsimonious; or they are ignorant and doubtful of its efficacy." There were sensible reasons for clinging to the old other than antipathy or irrationality.[87]

Chapter 4

Intimate Climates

Sydenham condemns the giddy practice of laying aside winter garments too early in the spring, and of exposing bodies over-heated to sudden chills. This practice, he affirms, has destroyed more than famine, pestilence, or the sword.

Alexander Sutherland, 1763[1]

Concerns with the medical consequences of daily life became a staple of the medical commentary on the Regency lifestyles of fashionable society and sedentary professions. In these commentaries, the medical body was determined by what and where it ate, what and how much it sheltered itself, what it used as a dress, and how long and where it slept or stayed in fresh air. From the practitioner's perspective, this was not a physiological so much as a "lived body."[2] This was also a "dispersed" body that required medical attention beyond that given to the pathologies of guts, flesh, and blood. Rather, the threat to such a body spread across space, seasons, and the everyday realm of life; what went under the skin was a footnote to what took place above and beyond it. Small-scale atmospheric pathologies and environmental change of the personal space presupposed a body affected by both culture and environment. Hygienists and physicians noted that health was a way of life to be sought in a knowledge of how to *place* and *move* oneself with regard to outside stress, be it sunlight, miasmas, frost, or graveyard smells. Just as physiology was a mimesis of exteriority, health was an achievement of being somewhere and assimilating the character of something. Its maintenance involved time-consuming routines—travel, recreation, hunting, suburban seclusion—and informed attitudes toward agents beyond the skin-bound body.

If how one felt was a sympathetic response to one's placement in nature and society, then there was no sense in speaking of an average body, or a normal or standard reaction to stimuli. There was no "typical" bodily reaction, and no fixed treatment even for the same symptoms. Rural populations showed more resistance to disease than their urban counterparts. Feeling well depended on the quality of one's surroundings even on the level of one's house, room, or bed. By the late eighteenth century, such discrimination in terms of scale found expression in medical thinking about clothing. In a manner similar to that shared by the analysts of domestic climates and ventilation, late eighteenth- and early nineteenth-century dress reformers staged a widely publicized renunciation of the customary (mis)uses of clothing materials. Their critiques were complicated by trends in fashion and the popularity of frequent dress change to accommodate weather and social occasion. Gradually, Regency practitioners took dress as a standard medical item and the flannel as the norm of health. In all such analyses, however, they stressed the limits of generalization. The employment, surroundings, diet, family dynamics, and material culture determined the looks, quality, and quantity of garment. What was healthy was not necessarily pleasant to touch. Experiments on thickness, texture, color, and hygroscopic and insulating properties of materials formed a basis for an emerging theory of dress. A series of trials on the properties of fabrics gave evidence for medical claims about the risk of wearing traditional fabrics and helped physicians in thinking about the pathological qualities of intimate microclimates.[3]

Such concerns were part of the understanding of health as a result of administration of all bodily activities insofar as they related to one's movements through daily surroundings. This understanding bestowed on medicine a greater authority in regulating conduct, making the eighteenth-century doctor the "diagnostician of social ills."[4] This authority extended to health advice in the context of household economy, but it also exerted influence over the ordinary articles of life, such as clothing, that had been traditionally regarded as a matter of taste, fashion, or economic circumstances. Medical practitioners such as Leeds pediatrician William Forster even used the term "non-necessaries" to capture the medical nature of daily activities and customary practices that differed from the traditional "nonnaturals" in that they were unnecessary for health but *could* cause a disease. Where one could not dispense with food, motion, sleep, and air, one *could* avoid doing things that were not necessary for healthy life: "[I]t is not necessary to live in foggy countries, or be confined to certain meats and drinks, or to sleep longer than we please."[5]

The non-necessaries were things "people expose themselves to by their own economy," including sudden changes of place, abrupt cessation in physical activity, and pathological situations due to neglect to adapt to temperature changes.[6] The term referred to the situations in which health could be affected by sundry tasks, anything from having a bath or breakfast, to standing in the workshop or walking with wet feet, to breathing in soggy inns or going to bed without a nightcap. Forster, however, emphasized that regardless of whether one looked at the nonnaturals or non-necessaries, both implied a particular process of disease causation. He used the Latin term "procatarctic" to described the occasional and thus unavoidable sources of diseases: "[I]f we narrowly enquire into every thing that has preceded our distemper, and if we regard symptoms and what is happening within us, then the knowledge and cure will come."[7] Such thinking announced a particular system of long-term pathology linked to the violation of elementary physiological functions. For Forster, such a pathology stemmed from the risks associated with the conduct of individuals and groups bearing marks of both cultural and physiological predispositions to do things one way rather than another. As a late-eighteenth-century medical writer noted, procatarctic causes could induce "a predisposition to a disease," derived from "what goes before."[8]

The fact that what caused a disease was what "went before" extended medical interest into virtually every direction of patient's spatial life and depth of his past. This made "procatarcta" hard to identify and allow the physician to identify their effects in specific cases. This problem was reflected in Boerhaave's discussions on the subject in his lectures on physics, in which he identified procatarctic causes as "remote" or "external" causes that could "excite the latent cause, that generated a new disease [...] So procatarctic (or occasional) causes put the predisposing causes in action." But when in other places he also called them "immediate," Boerhaave might have confused the reader by suggesting that an immediate cause could excite a predisposing one.[9] Regardless, the term was eventually used to describe the actions, situations, and exterior influences responsible for what is best thought as a meta-causation of disease. A procatarctic cause could be any agency that damaged the weakest link in the person's immunity, an injury that could trigger the action of an exciting agent such as cold or heat.[10]

Heat and cold, in particular, played a central role in medical thinking about what "went before." Understanding the presymptomatic errors leading to diseases such as the common cold could help diagnose a disorder and define a cure. Many things were important to recall: what the sufferer wore, drank, and exposed herself to. In this context, the fear of cold and the so-called "checked perspiration" produced by far the largest portion of medical advice in which prosaic undertakings acquired medical urgency if they violated the bodily heat management. The eighteenth-century idea that exposure to cold could account for bad health stemmed from the work of medical statists, beginning with the Paduan physician Sanctorius in early seventeenth century. Sanctorius and later researchers found that the *perspiration insensibilis*—insensible perspiration—contributed to more weight loss than sweat, urine, and stools combined. The "unobstructed insensible perspiration" became a regulator of bodily fluids and its vulnerability to cold and damp.[11] Checked perspiration could be occasioned in various ways. Rising from bed into a cool room without a gown was as dangerous as exchanging the sultriness of the ballroom for an outside frost. Idling in cold air was daring; playing with the clouds of mouth vapor was a gamble. Riding unbuttoned across humid grassland forced the body to absorb the vapor and jeopardize internal heat distribution. Leaving bedroom windows open at night was wrong even in sultry weather, as was reading or working near a window in a room in which door was ajar.[12] Venturing hatless, resting under the sun's direct rays, wearing wet stockings or shoes, walking barefoot on hardwood floors—all of this took a bite out of one's health.[13]

Medical statistics did much to shape the medical ideas about dress. For example, errors in clothing were held responsible for "catching cold." The name was given to symptoms such as sensations of chill, shivering, fever, and sneezing, and explained as a result of "exposing the body to windy and rainy weather, sleeping in the open air, especially in the evening, going by water, suddenly passing from a warm to a cold state, drinking cold water, changing the apparel, or living under ground."[14] Such "non-necessary" practices closed the pores and halted the flow of fine perspirable fluids. Results could be blight and ophthalmia, swelling of lips and nostrils, broncholele, and inflammation of the neck called a "crick." Ears could suffer as badly.[15] Conditions like these were benign when treated early, but if left untreated they threatened to cause fevers, nervous distempers, and catarrhs. Palsy was next to incurable when contracted during the winter, when the "frigorific particles" pushed through skin coverlets and forced bodily fluids to retreat inward. Stomach gripes, indigestion, and the wind resulted as whatever couldn't get out through the skin went back into the intestines and worked its way out via alternative routes.[16] For the affected, the simplest prevention was to keep warm near a fireplace and, in extreme cases, apply the method of "roasting" by "prodigious load of bed clothing" to bring on the "sweating crisis" and speed up recovery.[17] Forster considered it self-evident that the harsh Yorkshire winters justified making oneself "warm, by heaping Clothes, and making our Chambers hot" during card games, assemblies, and lectures. The cold, desiccating northeasterner improved appetite and called for fatter foods, stronger liquors, and brisker walking. Even longer night walking—normally a liability—helped to improve the circulation of blood. It made sense to suffer mild sweating during times in which "the Passions should incline to Mirth and Jollity, which move the Blood briskly on; and for this reason, Assemblies, Comedies, and Gaming do usually go forward."[18]

The Ascent of Flannel

Nothing controlled moderate sweating better than healthy clothing. By the second half of the eighteenth century, the specter of obstructed perspiration made clothing a subject of medical interest and led some medics and hygienists to inquire into the hygroscopic, conductive, and weatherproof qualities of fabrics. Whether a fabric helped or stopped the skin's release of invisible sweat was the key concern in all analyses within a new medical emphasis on the "clothed body."[19] Practitioners portrayed dress as an element of health, rather than a mere necessity or a matter of taste and status. "To this head of the non-necessaries," explained Forster, "clothing may be referred, which guards us from the injuries of the air; it ought to be proportioned to perspiration, which we must keep as near as we can to a stated quantity; wherefore in this country in which there are sharp differences between winter and summer, nothing deceives those with fever than the mistake of not putting on their winter garments soon enough, and of leaving them off too early."[20]

Early in the eighteenth century, most clothing was produced by the English woolen industry. People wore wool and flannel in combination with other fabrics, such as cotton (corduroy, velvet) and linen (as in fustian). In Defoe's description, the everyman's dress was made of cloth from Yorkshire, lined with shalloon from Berkshire; his breeches, drugget from Devizes; his waistcoat, calimanco from Norwich.[21] During the time of his writing, Indian chintz and painted cotton also began to enjoy popularity, which, being increasingly cheaper, led to a significant drop in sales of traditional woolen worsteds: "[T]he calicoes now painted in England are so very cheap and so much the fashion that persons of all qualities and degrees clothe themselves and furnish their houses in great measure with these."[22]

Wool profits additionally suffered because wealthier ranks sought distinction through quality, so that, for example, women of rank rarely wore straight wool except for warmth or in riding habits.[23] As a result, in 1721, the reaction of domestic wool industry led to excise duties on cottons. And while the act prohibited the import, sale, and wearing of all printed cottons, it had a relatively modest effect, because calico printers evaded the ban by printing on linen, which had not been covered by the legislation. The linen industry saw soaring fortunes as its sales spread downward on the economic scale. Linen shifts worn in the place of woolen ones were "a matter of neatness, comparatively modern,"[24] inspiring a Russian traveler in London in 1790 to note an "unusual cleanliness and tidiness in the dress even of the simplest people." The trend toward a widespread consumption of nonwoolen clothing accelerated after the prohibition of printed cotton ended in 1774. From this date, cotton, linen, silk, and muslin began to enjoy universal favor across all social ranks. The demand for low-maintenance items like chintz gowns boosted cotton manufacturing's fashion status while also trickling down to the less affluent. In the eyes of the public, even the finest worsted could not match the appeal of the flowing lines of soft-clinging muslins, thin silks of underwear slips, and plain cotton shifts. While wool did retain presence in common dress such as waistcoats, petticoats, and gowns, there was little doubt that traditional flannel sank to the level of utility and gauche winter gear. Cotton and lighter materials emerged as the top dogs of the late-Enlightenment sartorial sphere.[25]

The woolen industry suffered considerably. The problem incensed a number of analysts over the fate of wool, leading to outbursts of agitation aimed at fomenting consumer nationalism in the face of high-quality Spanish wool or cheaper imports from Ireland.[26] Influential individuals in the provinces organized promotional "Stuff Balls" for the encouragement of woolen manufactures. Lord Heathfield, a political economist and consummate military commander, led a campaign to make the western shores of Scotland a location for manufacturing fleecy hosiery.[27] Impromptu analysts of the relative merits of wool versus flax advocated government's involvement in the protection of the wool interest. Tracts such as Francis Moore's *The Contrast or a Comparison between Woolen Linen and Silk Manufactures Showing the Utility of Each Both in a National and Commercial View* (1782) assessed the material's place in English income through trade with the continent. Pamphlets on the moral and political economies of wool brought home the intensity of the depression in the trade, while the Society for the Improvement of British Wool and the Highland

Society of Scotland presented the issue as being of highest national importance. By the Regency decades, the polemics yielded to the authoritative digests of the "wool question" signed by names such as the famous stockbreeder Robert Bakewell and wool-stapler John Luccock.[28]

Yet while the crisis of the industry caused acrimony and argument, it also tended to bestow self-righteousness on those who defended wool on hygienic and medical grounds. By the time the wool question hit the national headlines, medical men established connections between clothing and health. The Reverend Roderick Mackenzie, minister and laird of Avoch, in his report for Sir John Sinclair on the state of health in the Scottish parish of Contin, noted that smallpox, measles, and rheumatism owed mostly to "the natives giving up the use of plaiding or coarse flannel next the skin. In the place of which they now wear linen."[29] This was also a cause attributed to the increase in respiratory disorders, influenza, and consumption among the manufacturing poor increasingly preferring cotton undergarments over those made of flannel and worsted. John Ferriar from Manchester Infirmary and one of the leading national authorities on fever and hospital practice thought that nothing short of a complete change in dress code would be able stop the rise of proletarian morbidity: "[I]t would be greatly to the advantage of the manufacturing poor, if the custom of wearing flannel next the skin were introduced among them. It would counteract, in many instances, the bad effect of cold and dampness, and would prevent many fevers and rheumatic affections."[30] Ferriar encouraged the establishment of clothing clubs aimed to promote general healthiness by a more frequent change of apparel. Patients admitted to the fever ward in Manchester infirmary were given the in-house flannel dress, which they were obliged to wear for the whole duration of recovery: woolen trousers and jacket for men and a wrapping gown and petticoat for women. At the time of Ferriar's writing, the decreasing use of flannel among gentlefolk generated a wide-ranging discussion on what this change meant for the future of both national agriculture and health. Critics emphasized the absence of medical grounds in repudiation of flannel clothes and linked the trend to ersatz gentility based on disposable income. William Cobbet found country dames and young ladies "foolishly ashamed of the clod-hopping shoes, quilted petticoats, and black worsted hose of their grandmothers."[31] Political economists argued that wool's role in the growth of British wealth owed to the providential geography of the country's pastures touched by the wet and balmy Gulf Stream. Doctors evoked the divine hand in giving the benefits of woolen wear to the manufacturing classes: Where animals defied bad weather with hair, fleece, and fur, men could do the same with flannel. Practitioners soon began a full-blown crusade to rehabilitate flannel as the fabric of choice in all socio-meteorological circumstances. They advertised its use in adverse weather and from the Shetlands to Egypt, but also in the ordinary course of life and among the ailing.[32]

The major problem of the campaigning was that it faced a deep-seated prejudice against flannel, which only increased with the revival of cotton and linen manufacturers during the 1780s. The culture of sensibility and refinement associated with light materials had created a negative stereotype of a rustic simpleton symbolizing unenlightened provincialism. The stereotype only reinforced the well-known arguments against flannel based on its tactile faults such as scratching and itching. Physicians often observed that, when worn next to the skin, its spicules scuffed the pores and roasted the wearer with additional sweating. "It does hurt to two for one that receives benefit from it," and is especially bad for weak, faint, and hectic people. If used continuously, flannel "relaxed" the body and by degrees stultified the skin fibers, hindering perspiration. It was hard to wash and, if rarely changed, it invited stench, rash, lice, and skin diseases. Its excessive use was said to cause disorders comparable to those effected by debauchery, grief, venery, and moist air, such as "crookedness, strait breasts, or any defluxion on the lungs, or any other of the viscera, distempers ill cured, and specifically cold in these circumstances."[33]

As early as 1737, Jeremiah Wainewright, a medical scholar exploring the mechanical operation of nonnaturals, gave an example of a consumptive woman from Sheffield who, although at the time able to perform routine tasks about the house, followed the advice of a physician and put on a flannel shift. As a result, "in two days time was confined to her bed, from whence she never rose without any other cause than wearing flannel."[34] Wainewright explained that flannel could cause problems to those to whom it is most often prescribed, "Weak, Faint and Hectic people." Good candidates for using the fabric were men of robust constitution who abstained from or were prevented from physical activity and thus could not "throw off the Remains and Dregs of a full and nourishing Diet."[35] Other contemporary critics noted that flannel's excessive production of near-body heat inhibited breathing, deepened anxiety, and caused nightmares and delirium fits. "Invalids and the people in health lift up every chink, damned to hot bed-chambers and self perspiration," were "often broiled to death" by using extra flannel against drafts.[36] The trust that the midcentury

A Hint to the Ladies – or a Visit from D.ʳ FLANNEL!!

Figure 8 Dr. Flannel suggests to a fashionable lady that she wear a flannel petticoat to keep her legs warm. I. Cruikshank, after G.M. Woodward (1807). Wellcome Trust Image Collection.

Englishman placed in flannel's insulation against endemic colds may be seen in the fact that even those who, like William Buchan, favored its preventive virtues, found the usage too excessive to be unmentioned: "[F]lannel is now worn by almost every young fellow. This custom is extremely preposterous. It not only makes them weak and effeminate, but renders flannel less useful at a time of life when it becomes more necessary. No young person ought to wear flannel, unless the rheumatism or some other disease renders it necessary."[37]

Physicians often lambasted popular fears behind the extreme measures that the "uneducated" public used to defend against cold. "Flannel next the skin" was a sartorial atavism known to weaken the body's natural resistance to cold and external stress. The practice was contrary to the benefits of exposure to bracing air and colder temperatures. The body required little artificial protection, demonstrated by the fact that the uncovered face suffered no injury regardless of weather. Clothing was a matter of cultural rather than natural consideration, because as it merely represented the need to hide shame or defect without concern for health. Warm garments thus might help those preparing to live in hot weather, "but not for us, who are to prepare our bodies for cold, which we cannot fly from in this country and should study how to resist it." Ancient Britons had gone about virtually naked and had better health than their descendants. Modern Canadians' disregard of shelter had similar effects.[38] One could even judge the level of civilization by counting layers of clothes, from the overclad brutes to the thinly covered Europeans and culminating in the upper-class taste: "[T]he gentry," wrote Hanway, "use the thinnest clothing."[39]

Eighteenth-century advocates of hardening made every effort to discredit flannel and other methods of keeping the body heated more than it required. Hardening was understood as both a *process* through which a body achieved a lifelong strength and immunity to disease, and a *method* of training whereby to reach such a goal. The popular poet hygienist John Armstrong wrote that "the body, moulded by the clime, endures / The equator heats and hyperborean frost / Except by habits foreign to its turn / Unwise you counteract its forming power. Rude at the first, the winter shocks you less / By long acquaintance: study then your sky / Form to its manners your obsequious frame / And learn to suffer what you cannot shun [...] a frame so steeled / Dreads not the cold, nor those ungenial blasts That breathe the tertain or fell rheumatism / The nerves so tempered never quit they tone, No chronic languors haunt such hardy breasts."[40] Such and similar attempts at somatic engineering took place in parallel with debates about acclimatization, degeneration, and species adaptation, all of which related to what might be termed a providential geography of racial vitality. Such debates raised the question of long-term adaptation of those living in nonnative climates, but they also addressed the issue of sartorial relativism and, especially, the English carelessness toward proper clothing. If French, Italians, and the Dutch wore waistcoats, why did the English avoid the practice? "I am fully persuaded," argued Forster, "that a little more conformity in this particular would be a means of rendering the lives of many persons, much more comfortable than they are; and would prevent many uneasiness, and diseases such as rheumatism stiffness, lameness, etc."[41]

The reason might be sought in the influence of John Locke, one of the most vocal defenders of light dress and exposure to the bracing British climate. Locke exerted a major influence on medical thinking about hardening when, in his *Thoughts Concerning Education*, he addressed upper-class mothers on methods of child rearing. Locke argued that mothers exaggerated their protectiveness through "cockering," mollycoddling, and overdressing. They used too many blankets and indulged in tight swaddling of newborns.[42] But when he went on to propose that children be brought up according to the methods practiced by the families of meaner status and knowledge, he sensed a controversy: "How fond mothers are like to receive this Doctrine, is not hard to foresee." It went contrary to the maternal instinct to keep the child uncuddled even though no one could say why the uncovered face would withstand more than the naked body. When an ancient Athenian asked a Scythian about how he survived walking naked in frost, the Scythian replied: "Think me—all Face." Bodies were made to endure anything as long as accustomed to do so from birth. Travelers through the Mediterranean, for example, observed that the Maltese had no difficulties in coping the worst heat of the island because they "harden the Bodies of their Children, and reconcile them to the Heat, by making them go stark naked, without Shirt, Drawers, or any Thing on their Heads" until they're ten. If this good practice was to be followed in the English climate—substituting heat for cold—the first move would be to reject overclothing, flannel pajamas, and nightcaps, as nothing "more exposes" to headaches, colds, and coughs than keeping the head warm. Those in doubt could be reminded of the cold-bathing customs of Germany, Holland, and the Scottish Highlanders that proved therapeutic especially in the case of children's health. Locke thought that, drawing on such examples, no gentlewomen should shrink from treating her children in the way poorer parents treat their simply because they do so out of poverty.[43]

Locke's recommendations echoed in the literature on thermal regimen and clothing, including cold bathing, sea bathing, hydrotherapy, ventilation, exercise, and climatological and travel therapies. Hygienists who followed his reasoning included household names like John Armstrong, William Buchan, William Heberden, and especially the influential William Cadogan and Michael Underwood, whose texts on the diseases of children featured extended reflections on the benefits of light clothes.[44] Cadogan, governor of the Bristol Foundling Hospital from 1753 and author the highly popular *Essay on the Nursing and Management of Children* (1748), discovered that the weight of flannel swaths, binds, and stays used on a newborn often equaled the weight of the child. It did not surprise him to find a healthy child become so "tender and chilly, [that] it cannot bear the external air, and if by any accident of a door or window left carelessly open too long, a refreshing breeze be admitted into the suffocating atmosphere of the lying-in bed-chamber, the child and Mother sometimes catch irrecoverable colds."[45] Even if things went well during this overheating, the child's health deteriorated the moment it was taken into the country to be reared in a "leaky House that lets in Wind and Rain from every Quarter." Cadogan's maxim in preventing these unnecessary traumas was that a newborn "cannot well be too cool and loose in it Dress," documented by the examples of infants "exposed and deserted" who could nevertheless live in relative health for several days. The examples of "exposures"—sad and heart wrenching as they might have been—demonstrated that the body could endure naked even when it appeared least capable of doing so.[46]

Campaigning for Warmth

Criticisms of hot garments did not go unchallenged. A more salubrious image of woolen clothing developed during the last three decades of the century among practitioners concerned with the emergence of neoclassical dress. In this period, urban middling classes began to favor a liberal cut, lighter materials, fewer layers, and freer bodily exposure. Among women of fashion, the style culminated in the early years of the nineteenth century as Directoire, Empire, or Regency style. Preferences shifted away from ruffled, buckled, laced, and brocaded elements toward the austere Empire silhouette of the 1800s, from hoops and panniers toward the informality modeled on casual undress. Corsets, reviled in the 1780s, loosened their grip, as the high-waist, nonchalant skirt, with deep open neckline and short sleeves, skimmed along the ever more visible outlines of the (female) body. Stylistic allusions toward egalitarian simplicity had been seen before, especially in the Arcadian style from the middle decades of the century, but these were perfected into a social statement by the Foxite Liberal Whigs during the 1790s and then by Beau Brummel, the unquestioned trendsetter of the unadorned appearance during the Regency.[47] His trousers, spotless tailoring, and the pristine standards of linen underwear became trademarks of modern taste and sexuality. Cleanliness and frequent dress change became *the* cultural virtue among the *haute couture* society and the more affluent middling sorts. The use and looks of underclothes underwent a change too: Women, especially after the 1790s, progressively reduced the total weight of their silk and linen underwear to only a couple of pounds.[48]

The press exploded with commentary on what exactly such developments meant for the market, gender roles, esthetics, and, importantly, health.[49] There was an outcry over the scantiness of female garb; the fabrics' transparency revealed more than it hid, leaving many confused about its advantage over the opaque outfit that had so far defined the public appearance. The ambivalence of scanty clothes presented a problem for conservative observers attempting to reconcile the clothed and naked conditions. Nakedness, in their view, represented the wretched condition of beggars and faraway tribes.[50] They were faced with the contradiction between the clothed condition as the body's public guise and nakedness as its private truth. Scanty robes placed the body in a problematic situation: "neither dressed nor exactly undressed, neither suitable for (re)presentation nor articulable as an Edenic condition."[51] The situation was even more problematic from a medical point of view. Moral critique immediately drew upon the fact that the new dress's sexual freedom was a climatic perversity: "Our ladies, alas!"—wrote Robert Thornton, author of the *Philosophy of Medicine* in five volumes—"are still compelled, whenever the enchantress waves her wand, to expose themselves, half undressed, to the fogs and frosts of our island."[52] The mortality rates from winter and sharp frost demonstrated that the cold weather and light dress made for a killing combination, literally. In casual commentaries like these, the frivolity of fashion engendered widespread micro-meteorological anxiety as doctors insisted that the delicate required warmth to be procured by means of flannel, fleecy hosiery, and the houshoeld economy of thermal comfort. The medical profession soon set about to define a new regimen in which flannel became a source of reinvented bodily welfare.

By the 1770s, the medical advantages of flannel were many and varied. Physicians argued that flannel created a pleasant micro-atmosphere—an "artificial, but uniform climate"—that kept the body away from cold and helped it release perspired matter faster than other fabrics. While it thus kept warmth without sweat, its qualities were especially praised in confirmed disorders in which its application could help those suffering from colic, diarrhea, rheumatism, and coughs. The usual way to use flannel was in the form of a shawl or cloth that could be applied to the afflicted part in two or more layers. If it was sprinkled with lavender or soap liniment, it could also soothe the inflamed throat if wrapped around it. It was also ideal for plasters, poultices, coverings, friction therapy, and resuscitation of hypothermic victims.[53] Physicians considered it far superior to linen in relieving the symptoms of rickets, because its "bibulous" quality facilitated a better uptake of both bodily moisture and external humidity.[54] This virtue was often referred to in the healing of remittent fevers during autumn or in humid spaces such as the Houses of Parliament.[55] It indeed looked as if the material had an unusual variety of uses, some of which looked closed to contrary. For example, supporters of hardening recommended flannel for keeping warmth among those "delicately brought up,"[56] while colonial practitioners said it could cool the body in even the sultriest of climates.[57] However one looked at it, flannel enjoyed "the great reputation [...] justly and universally acquired for its efficacy, and the very extensive sale in consequence thereof."[58]

But the growing popularity of lighter materials and stylistic freedom caused friction between the flannelists and the fashionable. One of the first extended critiques on the dangers of new styles was penned by Thomas Hayes, a London practitioner, who in his *Serious Address on the Dangerous Consequences of Neglecting Common Coughs and Colds* (1783) outlined a pathology of modern dress code in what might be seen as a template of the medicalization of social mores and material culture during the late Enlightenment.[59] Hayes's reflections on the consequences of libertine fashions looked into the abuse of "non-necessaries," especially the uses of female garb. During his out-of-town practice, he noticed the extent—not seen before—to which even countrywomen risked attacks of cold when, pandering to latest city vogues, they wore unnaturally short stays, great hoops, petticoats above knees without any undercoverings and woolen drawers. "It is matter of some surprise, that delicate as they really are, more mischief does not accrue from such modes of dressing." Such clothes allowed for the inner circulation of air, creating next-to-skin drafts that compromised perspiration. Adding to these threats were the more frequent daily changes in the amount and thickness of dress, which exposed the skin to sharp thermal amplitudes. Women of fashion spent mornings guarding their necks and chests, wrapped in gowns, caps, and handkerchiefs, abandoning them later in the day and venturing "half naked in the street" or tramping from playhouse to dinner party without care for wind eddies, fog, or rain. Neither was it obvious what rationale would lead young men, buttoned up during the day in accounting houses, to cast off their overcoats just before going out: "[S]o indiscreet is pride, that you seldom see them in a great coat when they are dressed for the evening, although they have been wearing it almost the whole day before [...] We could wish the morals of the people were such as not to require its so frequent exhibition; but as we cannot expect to reform the age, we think it our duty to recommend warm clothing, whist they are requiring its specific virtues, and it may not do more injury than good."[60]

Hayes criticized unseemly exposure that mixed promiscuity with aerial risks.[61] He attacked the sexualization of dress, the body's visual accessibility, and shrinking social proxemics. A "random," titillating disclosure of the lower leg, cleavage, and nipple violated both comfort and health. They undermined the thermoregulation of good old fabrics and hardy chemise. Culture, not nature, made the fop ill, observed daily in the lives of coachmen and farm workers whose health "we attribute to their going so warmly cloathed as they do."[62] Hayes censured the extreme separation of looks and use, ornament and substance, decorous chintz and hearty flannel. Chic nudity aside, women and men still had legitimate concerns over the variability of weather and how, if at all, to keep it away by means of dress. Was it better to keep the same flannel waistcoat until it rebelled against the nose, or was regular, perhaps frequent, change necessary? If so, when and how often? What was the trigger of change? How convenient, how expensive, and how symbolic of status was this practice? Was it wise, as Hayes observed, to know that "the human body becomes necessarily exposed" during the clothes' change?[63]

Most of the time, medical commentators on dress could give only haphazard answers to these questions. In 1787, however, the versatile Benjamin Thompson performed a series of experiments designed to determine the relative hygroscopic properties of several species of textiles. Thompson, a natural philosopher and philanthropist with a rich military, political, and scientific career, had already been involved in the study of electricity, heat, and friction, which during the 1790s resulted in the construction of the "Rumford Stove." In his 1787 communication to the Royal Society of London, Thompson explained that the method he used was to hang textile samples in a damp cellar for seventy-two hours and then measure their weight. The heavier the sample, the greater was its hygroscopic potential to absorb sweat. Somewhat to his surprise, he repeatedly found that wool weighed the most and cotton the least. But why would materials like cotton and silk, known for their good water absorbency, weigh less when surrounded by water vapor? Thompson argued that it was important to dissociate measurements with the subjective feeling of cold and wetness when in direct contact with skin. In the case of linen, for example, the feeling of dampness on touch was not due to the actual water the piece contained, but to the "ease with which that substance parts with the water it contains." Similarly, a body seemed to be hot to the touch as a result of "its parting freely with its heat, while another body, which is actually at the same temperature, but withholds its heat with greater obstinacy, affects the sense of feeling much less violently."[64] This explained the common knowledge that wool, however coarse, soaked up sweat regardless of its natural warmth. For if wool's hygroscopic quality resulted from its warmth, the same effect could be achieved with multiple layers of any other fabric. This, however, could not be accomplished using cotton, linen, and silk. But it could with wool, which transmitted body vapor to the surface of clothes and from there into the surrounding drier air. "The loss of this watery vapour, which the flannel sustains on the one side, by evaporation, being immediately restored from the other, in consequence of the strong attraction between the flannel and this vapour, the pores of skin are disencumbered, and they are continually surrounded by a dry, warm, salubrious atmosphere."[65]

While these results and reasoning alerted the medical profession of the benefits of wool worn next the skin, Thompson was fully aware of the stigma attached to the fabric by metrosexual elites. He nevertheless teased the reader by admitting that he himself had known "greater luxury than the comfortable sensation which arises from wearing it, especially after one is a little accustomed to it."[66] Following the recommendation of the physician Richard Jebb, Thompson himself wore flannel in the hottest of climates without any trouble, itching, or feeling drained by its "roasting" property. If anything, he found that its breathability let him stay cool even during physical activity. "It is the warm bath of a perspiration confined by a linen shirt, wet with sweat, which renders the summer heats of southern climates so insupportable, but flannel promotes perspiration and favours its evaporation; and evaporation, as is well known, produces positive cold." "Trifling as they may appear," Thompson admitted of his experiences and experiments, they were the most important he "[had] ever made."[67]

This was big news in the flannel community. Thompson's authority gave a convincing solution to medical guesswork. Experimentation on heat conduction could both rescue the economic providentialism of "Stuff Balls," and vindicate the medical and moral arguments of the pro-flannel traditionalists and critics of urban mores. Thompson's findings appeared in Adair's tract on fashionable diseases, in which Adair offered detailed recommendations for flannel's applications, even a seasonal timetabling (a waistcoat over shirt during September, and next to the skin from October to May). The legs in particular—exposed to the draftiest strata in households warmed by open fires—required heavier petticoats, woolen drawers, and worsted stockings. This might have not been unreasonable anyway, given Adair's recommended indoor temperature of a meager thirteen degrees Celsius (67 degrees Fahrenheit)![68] Even Buchan, a reluctant supporter at first, used Thompson's observations in his "witness report" on behalf of George Holland's fleece manufacturer.[69] Buchan personally vouched that Holland's products had a outstanding warmth-to-weight ratio, giving them an advantage over all other candidates: "Had Sir Benjamin known how far your Manufacture exceeds flannel, both in agreeableness and use, he would have been still more lavish in praise of it. Indeed it is hardly possible to say much in favour of a medicament of such extensive powers, and which is so pleasant in the application."[70]

Thompson's results inspired more extended medical analysis of clothing, such as Walter Vaughan's *An Essay, Philosophical and Medical, Concerning Modern Clothing*. The book was published in 1792, a year that saw the fall of the male ruffle and the launch of Francophile neoclassicism in Britain. Vaughan documented the urgency of the subject by attempting a Newtonian synthesis of sartorial folklore. His study was to look into *"the causes of dress"* and explore the relationship between modern apparel and "inability, disease, and death."[71] He canonized medicine as the only guide in the everyday rationality of clothing practices, arguing: "So much it is incumbent on physician to study and inculcate what most conduces to health," even if such an ambition had earlier been undermined by custom and prejudice. His work was thus less a contribution to the genre of medical self-help, and more a paradigmatic example of what might be termed a medicine of the mundane: "[I]t is an observation easily made every day that man considers those things the least of all, which from their relation to him he ought to consider the most; people generally are reluctant to avoid what they don't consider to be wrong, so it is necessary that they first be convinced that this is the case."[72] Vaughan had no doubt that pedagogic "inculcation" should benefit humanity, whether or not, as a consequence, it would also bestow a medico-meteorological meaning onto the spaces, items, and practices normally outside the physician's remit.

The book was a result of personal observations during his continental travel with J.E. Smith, then president of the Linnaean society. Vaughan denied that his dislike for metropolitan fashions in any way influenced his professional criticism. He was not arguing against fashion as such, but against its erroneous medical foundations: Votaries of fashion should not fear that his had any preconceived "imposition" about what was essentially an aesthetic subject. He was instead going to prove that "the Notions which we have of the proportions and beauty of the human body are arbitrary and fanciful," rather than based on the notion that clothing served to defend from the inclemency of weather and hide private parts in the interest of decency. But "by being worn agreeably to the present fashion, they incapacitate us for our duties as members of society, create distress and disease, and are in truth the most insidious instruments of self-destruction."[73] Such tribulations relied on a providential correspondence between bodily form and function, as Vaughan argued that whatever altered any part of the body was by definition an act of degeneration. Vaughan challenged Comte du Buffon's rule that fashion was rational as long as it concealed defects: If that were the case, every fashion was rational.[74]

There were examples to prove this point. Women's clothes were often made too small; hands suffered pressure by the sleeves of gowns and coats. Especially painful were partial pressures from the hem of sleeves, bracelets, garters, shirt wristbands, or the elastic band and buckle worn to fasten gloves. Plentiful wadding was a source of compression that could make a dress a ligature. The bulk of a person also played a role: large bodily frames strained fabric and made diets and fasts measures that only increased the health risk by reducing the lipoid shell that guarded the body from colds, coughs, and consumption. Neither health nor beauty benefited: "[D]o not the breasts of women so white, exuberant and lovely by nature, become disgustingly pendulous, olive colored and flabby when they lose their fat? [...] I wish I could persuade my fair Country Women to bear with patience that complement of fat allotted them by Providence."[75] For the less fatty, especially of delicate constitution or undergoing a recovery, the ideal apparel was woolen, made of "most natural and the most wholesome" material, maintained a uniform temperature of the intimate air layers. But a person dancing in a linen undershirt had more to do to manage the sweat and stench: "[W]hat will be his sensations! What his Risqué! His linen will be soaked in Sweat, and like every Thing excrementiously disgustingly stinking, he will feel cold and shiver, his teeth will chatter, and it is a thousand to one but he catch Cold—a Hundred to One but his Lungs become inflamed."[76] Equal "risqué" attended the vogue of frequent changes of dress that Vaughan characterized as a perverse social convention in a climate where adapting to weather was next to impossible. "I am hurt when people in ease and affluence tell me that clothes should be changed as often as the weather changes; just as if they had only the care of themselves at heart. For poverty will always preclude the laboring poor from the advantages of so frequent a change, if it is necessary."[77]

Keeping a thermal calm next to the body was the main argument for uniformity in dress. Again, wool fabrics suited this purpose best in virtue of slow heat conductivity, allaying the amplitudes of climate and working conditions, filtering out miasmas, and preventing infection. These assumptions had found a supporter in Anthony Florian Madinger Willich, whose comments on the patient's hypersensitivity were discussed in Chapter One. Willich understood the merits of woolen garments not only among those "too susceptible of the impressions of the atmosphere," but also among those suffering from occupational exposures. His *Lecture on Diet of Regimen* underscored the universality of flannel as a shield against the labor atmospheres: The manufacturers in Colebrook Dale foundries and ironworkers in Kettley wore it in the hottest factory rooms, as did members of other professions "exposed to all kinds of weather," including British soldiers, postmen, and farmers. Flannel required little change in the under dress and was a good substitute for the deficient upper dress. It deserved the attention of officials in charge of the indigent, orphans, and poorhouses as a preventative technology for the vulnerable and underfed.[78]

Willich's list of evils of dress were already known, but his attention to hazards were finely grained, the analysis more discriminating and diagnostics more worrying than ever before. His concern did not stop with the physical features of the fabric. It extended to issues such as a garment's shape, length, thickness, electrical properties, and seasonality. Even the color could have a pathological quality: White reflected the light; crimson and scarlet caused a vision blur, as did the glare from steel buttons, gold, and silver lace. Minuteness of this sort is not seen in earlier discussions. Healthy clothes were to be pliable, thermostatic, and movement-friendly. Shirtmakers were to advise to dispense with narrow collars and tight wristbands. Tight neck-cloths, cravats, ribands, buttoned-up shirts, and necklaces caused fainting, stupor, and headaches. Scotch bonnet, Dutch cap, or Turk's turban—as well as other hats, wigs, and nightcaps—broiled the scalp without benefits for the brain. Indeed, asked Willich, what leads people to "starve their feet while enclose their heads in an artificial stove" or a "vapor bath."[79] Similarly, constricting leather breeches were inferior to looser pantaloons of elastic woolen cloth. Sweat-absorbent cotton stockings could not match woolen socks worn under waterproof shoes and supported by "thick cork soles for the winter, or with elastic socks of horse hair." Warming the feet and cooling the head, Willich worked to diminish the vertical gradients compromising the homogenous heating of the "unclothed" body.[80]

Such claims insisted on dermatological benefits of wool's coarse texture, pronouncing flannel as *the* intimate "technology" of comfort, health, and positive pleasure.[81] For the delicate and the elderly, it might be a second skin: A "mathematical valetudinarian" explained that "in the ration that my body would be uncomfortable without my skin, would my skin be without my flannel waistcoat."[82] Despite its purported "relaxing" qualities, wool's use by the public had actually increased by the early nineteenth century: "[T]he extreme variableness of the climate of Britain," noted the Earl of Dundoland, "with the advice of many eminent physicians, has induced a considerable proportion of the inhabitants of this country, to wear flannel and fleecy hosiery next the skin."[83] By 1800, hosiery production was underway in Godalming in Surrey, and there were plans to convert one of Aberdeen's colleges into a facility for the treatment of Scotch wool into fleece, from where exports to Russia looked set to rise, the product being "so well adapted to northern latitudes."[84] Randolph Nott, hosier and dealer "in all sorts of flannels" from Newgate Street, announced that his nine-times-dyed blue flannel had been praised by physicians for its beneficial effects on gout and rheumatism. Nott noted the city's flourishing supply, which impinged on the business of soda-based washing powders, sold to keep odors at a tolerable level.[85] Flannel's popularity continued well into the nineteenth century.[86]

When in 1793 public pressure increased for a subscription for worsted stockings and flannel caps for British soldiers in the European theater of war, wool had seemingly fulfilled its providential purpose in serving its homeland's fight for freedom from political adventurism.[87] Enlarging upon the plan of an Edinburgh committee for flannel waistcoats—"to give our brave countrymen on the Continent every comfort to be derived from warm clothing during the winter season"—the *Times* launched an appeal to keep British feet warm in fighting the Gaul and to prevent ills from "obstructed perspiration." It took less than a week for a charity in Charing Cross to collect more than 3,600 pairs of stockings, 2,400 caps, and 58 guineas. Randolph Nott had a great day, and thanked the public for patronizing his business for the best of causes.[88] Not everyone joined the hype, however. In the pamphlet collection titled *Liberty Scraps*, the piece "On the Flannel Waistcoat Subscription" made flannel an emblem of national peevishness and lack of political vigor:

> When Britons fought in freedome's glorious cause,
> They fear'd no cold; they gain'd the world's applause;
> The case revers'd—warm waistcoats they require,
> And droop, uncheered by freedom's sacred fire:
> Not so the French—they feel the powerful glow;
> Nor heed the piercing frost nor driving snow;
> They ask no flannels, but camp repair,
> And sweep all conquering, tho with backsides bare.[89]

Likewise, not every physician approved of the warmth. Thomas Beddoes, a Bristol celebrity social critic with credentials in pneumatic therapy, harangued against flannel in his *Manual of Health*, which he thought of as a medical chronography to guide patients "through the seasons." But he could not miss a chance to stalk a bigger prey. His real targets were the "degenerate members of civilized society" made up of self-regarding elites infamous for their obsession with comfort and overly worried about aerial surroundings.[90] He was aware of flannel's benefits in protecting those "exposed to the variation of the atmosphere," but was at pains to "understand why any one who is master of a comfortable bed should wear flannel next the skin in the night. The practice is contrary to common sense," because flannel boiled the skin into feebleness. It should be replaced by calico, which was ideal for the sedentary and indoor-bound. The chronic complaint of chilliness among deskbound professionals and women of fashion was, in his view, nothing but an artifice of overclothing. Fleecy hosiery, despite the propaganda, "has proved to be injurious, constantly provoking excessive sweating. The wearers seem to be exhausted, and often like ghosts [...] as if just risen from the bed of sickness."[91] Beddoes was impatient with any other use of fleece than one directed to cure numbness and rheumatic palsy. He inveighed against the delusional piling of layers to suppress "chilliness" and the extra "receptivity" to changes in the air. Was it not, in truth, absurd to believe, that a delicacy, which shrank as a result of variations of weather, could be removed by heaping additional coverings on the body? The point was brought home in an account of one of his patients:

[W]hat I have to complain of is extreme suffering from cold. [Another doctor] wishes me to increase my quantity of cloathing; I have strongly resisted the suggestion from conviction of its inefficacy. I am wrapped in fleecy hosiery, have thick understockings and silk above. I have, besides, fleecy hosiery drawers, two petticoats, and my gown. I wear long sleeves with gloves under, and a very thick double muslin handkerchief. On going out I put on a cloth pelisse, trimmed with fur close up to the throat, a muff, and a pair of very strong thick shoes over my others. In the house I usually wear warm list shoes also over my others. I have been cloathed in this manner for the last two years, and my chilliness is certainly increased [...] When I am sitting in a room, I often feel the cold occasioned by opening of a door, as plainly and as immediately in my bowels as I do on my face—I wish to know whether you think *more* clothing can be beneficial.[92]

Figure 9 A parody of chronic sensation of cold. *The Man That Couldn't Get Warm* (1801). Wellcome Trust Image Collection.

Wool Incarnate

Some of the concerns expressed by Beddoes's patient could be understood in the context of contemporary interest in the role of cold in inflammatory disorders. Fear over partial exposure to winter air was widespread among those inclined to inactivity or cursed by chronic disorders of the mucous membrane.[93] By the mid-nineteenth century, public and professional commentary on the issue of cold in relation to clothing found an expression in the study called *Thermal Comfort*.[94] Published in 1843, the work was the brainchild of Sir George Lefevre, fellow of the Royal College and physician to Polish nobles and British diplomats in St Petersburg.[95] Affected by fragile health, Lefevre initially traveled to the Mediterranean and Pau, but ended with practice in Russia and England. Personal experience made him a champion of warmth as a health preservative. Encounters with the living conditions of the Russians convinced him of the invalidity of the ideas about hardening and acclimatization. Contrary to what had been regularly assumed about the correspondence between bodily frame and local environment, providence showed little mercy in safeguarding native health from native air. Residents of subpolar nations suffered from chilliness much more than southerners if placed in the same climate; if they did not suffer from it, it was because they stayed inside. With Lefevre, the notion that delicate bodies needed steeling is gone without a trace. What remained was a quest to create a social space of warmth and equability: "Heat breaks no bones." Those who argued that the sensation of comfort was too subjective to be prescribed as a prophylactic, Lefevre reminded them that personal feeling was the key to thermal medicine. Aren't the senses what prevent us from exposure? "I repeat, it is not the man who resists the cold. It's the man's clothing, it is the provision he makes against the cold. Let him imprudently traverse the courtyards without his hat, and with no clothing beyond what he wears in the warm halls, and then what awaits him is pleurisy, dropsy, and slow death." Unlike the English, the Russians respected the risk when, in their thick fur layers, they carried, in compact form, the heat made in crevice-free cottages, and were content with what English pundits of hardening scoffed at as debilitating "vapor baths."[96]

With Lefevre we see a temporary endgame of flannel ascendancy. The moment was temporary because the rational clothing campaign as enunciated by people like Gustav Jaeger was just around the corner.[97] But Lefevre's study of costumes and climates opened a window toward an anthropology of weather as the source of personal as well as national ideas of honesty, energy, and bravado. Lockean apologists of hardening silenced, the warmth as an index of pusillanimity remained a belief in lands where the sun presented little threat. In England, the much commented upon resistance to technologies of household heating might have been an indirect refusal to accept a meteorological reality that was far too unflattering to be canonized in the flannel dress. As Lefevre made clear, the penchant for warm clothing expressed a high-minded consumerism instigated by know-how of the clothing industry. This know-how was, in turn, a solution to the anxieties over the social and somatic abuses of the "non-necessaries."

But for all of the wording that stressed inspissations, perspiration, conductivity, hygroscopy, miliary disorders, debility, etc., the ultimate criteria for judging the relevance of the solution eventually had to do with firsthand practice. Personal experience in the medicine of daily clothing mattered more than an occasional self-admission would have us believe, partly because firsthand experience was the basis on which layperson and professional views could meet without conflict. Patient and doctor shared sartorial conventions that allowed them to talk about the subject as real-time wearers, not as unequal participants in the esoteric skills of diagnosis. These implications are worth taking on board because virtually all of the contributors to the topic had been experienced flannelists who formed their opinions through self-experimenting. Like the great Robert Boyle, whom they all regarded as a shining example of the need to adapt dress to seasons, the leading medical advisors on apparel found it necessary to try to incarnate their science into toilette: "I have always seemed to myself to find advantage in changing flannel for calico," Beddoes said in his attack on flannel pajamas, "for any one can keep up his warmth by a due accumulation of bed clothes."[98] Similarly, Wainewrigth, Thompson, and Willich used their personal practices as the source for their views on fabrics and health. Thornton and Adair knew from experience that the waistcoat should be changed once a week and, in warmer weather, worn over linen and completely abandoned by the summer.[99] Jonas Hanway preferred warm clothing and, in later years, "was ordered by his physicians to the south of France: but some very urgent business calling him to visit Holland, in the moist air of that country he recovered and remarks that the Dutch are less subject to this dreadful calamity, which he attributes partly to the air, and partly to their warm method of clothing." Hayes fought his infirmity with flannel, and Vaughan wore it while exercising; Thornton used fleecy hosiery in hot weather. None reported inconvenience; all advocated flannel.

The fact that the medicalization of clothing can be reconstructed from autobiographic experience should not detract us from seeing the process as a medical program with social consequences. Medical theory of clothing fetishized personal idiosyncrasy into rational conduct.[100] But the autobiographical sources of dress regulation were important also to the extent that they stressed the scaled-down "environmentalist" attitudes during the late Hanoverian era. These attitudes showed how nonnatural and non-necessary behavior shaped and reorganized, compulsively and with self-regard, the daily *lebenswelt* of the English middle classes between the 1770s and the 1830s. Actions previously treated as routine—from where one slept to what one wore, from how one moved in and out of the house to how often one opened the windows—had during this period acquired a status that propelled them into mainstream medical theory. Yet this medicalization was not a result of the transference of military rigmaroles onto civic life, as Foucault assumed, nor a bid for hegemony on the part of professional opinion, as sociologists of health continue to stress.[101] It was rather a way of defining identity in a society that derived its notion of status from a "positioning" oneself with respect to the ideals of public and domestic space. The economic, somatic, and cultural sites of such maintenance should be seen as the grounds of social orbit as related to individual's comfort and well-being.

Chapter 5

The Choice of Air

Sire Henry: Yes—I can't bear to be stationary, we none of us can, and I verily believe that to breathe the same air for twenty four hours would be the death to me—it would be the death to all of us, would not it Jack? Of this, at least, I am sure, that I should have no appetite, and what would be life without appetite?

Edward Morris, 1799[1]

M. Do you think we live in a purer air than the gentry?
T: Their habitations are less confined, and they go abroad, or stay at home, as the season, or softness of the weather invites; and consequently have a choice of air, if they will use it.
M: If they expose themselves wantonly in the midnight revels, they put themselves more upon a level, with regard to the unavoidable evils that we suffer. Would it not be more happy for them to face a winter's sky, being fortified by proper garments?

Jonas Hanway, 1774[2]

We have seen how the economic status and pecuniary arguments underlay the aerial doctrines of indoors, clothing, and, more generally, the theory of nervous sensibility to atmospheric powers and climate change. Socioeconomic considerations, as Hanway observed, were equally powerful in driving social and medical commentary on what universally came to be dubbed as the "choice if air." This was not surprising. Money could buy better air as the affluent began to free themselves from a fixed abode. Permanency of labor entailed permanency of abode just as economic dependency entailed vulnerability to local disease. Some could choose to avoid risk; others were forced to adapt. Renaissance nobility headed for the hillsides to evade the miasma; the poor remained in the valley to cope with the stench.[3]

During the eighteenth century, moving to better air became a more common practice and, eventually, an orthodoxy. This might partly be due to the popularity of the European Grand Tour. Never before had medical advice been so favorable to a practice that, by the early 1800s, made "choosing the air" one of the most appealing methods of therapy.

Views, however, differed on benefits, feasibility, and hazards. Early authorities like Arbuthnot identified the lung sufferer as the chief beneficiary, but were skeptical as to the practicability of the therapy: "III mankind, by Reason and their locomotive Faculty, have in some measure the Power of defending themselves from the Injuries of the Air, but few have the Choice of the Air in which they live."[4] A mid-century Lincolnshire topographer wrote that only those "Persons as have no Call to reside in these dangerous Climates, [can] depart with speed, if they would avoid Fluxes, Quinseys, Pleurisies, Fevers, Agues, Consumptions, and other violent and almost incurable Diseases."[5] But even for those *with* a call, not everything was lost. A more granular understanding of living milieus left choices for stationary populations. Eighteenth-century asthmatics were thus admonished to seek warmth and viatic purity by short-distance commuting: Those "obliged to be in town all day ought, at least, to sleep out of it."[6] Recommendations like these made sense within what might be described as a practice-driven fragmentation of space that enabled one to consider the quality of atmospheric life in terms of anything from property prices to window-opening methods. The very possibility of choosing one's air implied a distribution of healthy properties across all spatial scales, from privies to gardens to neighborhoods and beyond. "Surely the dirty Road that lies thro' the Town, and the Brick-kilns about it, cannot but be offensive to those who pretend to be nice in the choice of their Air."[7] Here, as elsewhere, the "objective" quality of local atmospheres depended on the economic and technological feasibility of alternatives.

It is not therefore surprising that one of the least feasible alternatives—long-term international travel with no guarantees of recovery—became one of the most popular. This it owed to its purported efficacy, even if efficacy might have been derived from exclusivity. We may see how early practitioners (if not victims) have considered it "unprovidential" that a method of cure requiring comprehensive financial and locomotive logistics should betray theirs and their families' hopes and tarnish trust in medical mentors. The length to which the English patient was prepared to go to immerse her constitution into better air seemed just about sufficient to legitimize the scientific basis of the cure. And while the travel/climate therapy offered an alternative to orthodox measures at home, its proponents never failed to acknowledge the extra effort, organization, and material circumstances that health travel necessitated. In so doing they laid the basis for a social emblematics that climatotherapy began to enjoy during the latter part of the enlightened century. By degrees, it became apparent that for wealthier invalids—those who could afford to lose faith in local practice and native climate—the very prospect of "change" exuded a prophylactic appeal unmatched by even the best-planned home treatment. Face-to-face consultation was the norm. It could be expected that such an interaction would have a theoretical correlate in constitutional medicine's stress on the idiosyncratic and the contingent. Proponents identified the conditions whose cure required rethinking, reordering, and avoiding the suspicious surroundings: A "change of object, change of air, change of things to contemplate" contributed "*something* in favor of the patient; and when it is considered that this *something* is not be found, but by this change; I apprehend that the propriety of the measure, is too evident, to require farther demonstration of proof."[8] Tautology notwithstanding, such arguments painted "the change of air" as an enlightened approach toward pulmonary health in which travel for warmth could also, at first vicariously, and then blatantly, be undertaken as if it were an item of conspicuous consumption. Even the earliest proponents of sea voyage therapy, among whom Ebenezer Gilchrist was most frequently cited, knew of the appeal of the practice chosen by those suffering from "a mere modish affectation of being ill."[9]

By the early nineteenth century, medical literature gave considerable attention to the "election of air" as a preservative of respiratory, gastric, and nervous health.[10] Practitioners addressing the strictly therapeutic (or preventive) role of the "choice of air" did so occasionally in the context of the "six things non-natural."[11] The "choice of air" and "the change of air" had a basis in neoclassical humoralism, nerve theory, and constitutional physiology. Yet while commentators made clear that choice of air was "the principal requisite to good health,"[12] it was a matter of considerable contention as to how, by whom, when, and where the method was to be used. Not all who were in real need of better air could go; not all who went needed to do so. For the time being, however, the profession was happy to leave the option on the margins of practice and consider it an accessory to normal practice. "Climate change" was something to be recommended when everything else had failed or when a recalcitrant ailment pressed the patient—and especially his relatives and friends—to seek unorthodoxy. But just what specific power resided in "better" air was a subject of debate if not embarrassment. Was there "a certain vivifying principle, necessary to life, of which we have no determinate notion"?[13] Did a change of air helped to heal because of a change in heat, reduced humidity, and the stimulating electrical charge of the atmosphere? Or should its power be attributed to alteration in light, landscape, or earthly emanation? It might also have had to do with intensity of and difference in leisure, exercise, lighter clothing, or healthier sleeping hours.

The initial appeal of health travel owed a great deal in particular to the failure of drugs—a view forcefully imprinted upon public and professionals by, among others, the medical heavyweights James Gregory and William Falconer. Gregory, a widely traveled, Leyden-educated professor at the Institutes of Medicine at the University of Edinburgh, considered British winters the "exciting cause" of chronic disease for which the physician had no remedy but to wait for the summer or *travel* to it.[14] In his 1774 dissertation *De Morbis Coeli Mutatione Medendis*, Gregory asserted that the most perilous qualities of British weather were its variability, cold, and moisture, especially with regard to pulmonary maladies, but not exclusively so. Cold weather constricted the blood vessels; moisture relaxed muscles and weakened the contraction of the heart and arteries, suppressing the blood into the inner body and creating therein a tendency to inflammation he termed "phlogistic diathesis." Obstructed perspiration lessened the nervous tone; sweat turned viscid and the skin rigid, lessening the "accuracy of sensation."[15] In the words of William Falconer, the Bath doctor whose patients included William Pitt and Horatio Nelson, a change of place was a necessity: "[I]t is obvious that the remote predisposing and exciting causes of many diseases arise from the influence of our climate [...] This may be obliterated by moving to another climate."[16] While this opinion might have seemed meaningful from the physiological point of view, it was less so when examined in relation to actual disorders and social constraints of sufferers. Herein lay a tacit social aesthetics of health travel. Gregory explicitly identified the city dweller as the most likely beneficiary, even though it went against common knowledge that such a person was least likely to run the risk of suffering from a phlogistic diathesis due to exposure to outdoor cold. If this paradox is anything to go by, the popularity of climatotherapy derived less from advice given on the basis of the geographies of British morbidity than in an understanding that its object was an overhaul of constitutional frailty, independent of any specific disease.

Gregory considered it proved that polished society was most "solicitous to elude the rigorous excesses of the weather." Overeating, overheating, overclothing, rebreathing, overprotective living, and other enfeebling trappings of sudden wealth made the metropolitan body unsupportive of frigorific annoyances. He reasoned that pathological conditions attending the pursuers of modern living required an antidotal treatment of both predisposing and exciting causes, not a cosmetic amelioration of the symptoms. "Affluenza" required introspection in a balmy air and a sylvan stroll in a far removed land: "[I] t is in cases then of this nature," wrote Gregory, "that they should seek another climate blest with the cheering aspect of a clear and serene sky...where the prevalence of pleasant and more agreeable manner and livelier scenes are calculated to soothe the feelings, to promote cheerfulness etc." Relapses brought on by a bad air could be halted by a planned travel "since their continuance in this island subjects them to perpetual exposure"[17] to British noxious atmosphere, seasonal changes, and urbanization. The virtues of air change were superior to the effects it might have on either mind or body, directed as they were toward a recuperation of stamina. This of course was a feat less dependent on the thermometric qualities of a resort than the advantages of locomotive, sensual, social, and other peripatetic diversions. Travel itself, with necessary changes of scene, language, company, vegetation, and diet, was sufficient to jump-start a languishing constitution. Travel, not air, was the real pulverizer of disease and climatotherapy was a method of planned oblivion. It moved sufferers from spaces that bore the symbolic indexes of their medical histories into an agoraphiliac innocence of spaces bearing no such associations. The change of air was a philosophy of disengagement that pursued health by distraction.

Doctors, however, were quick to realize that the continental journey was not the sole source of successful treatment. Traveling was coincidental, not instrumental to recovery. If tolerable conditions could be found locally, so much the better: "[I]f the climate cannot conveniently be changed, we always have it in our power to alter the nature, and qualities of that particular atmosphere in which patients breathe; or, in other words, we may accommodated the nature of the air to the nature of that season which is known [sic] to be most healthy."[18] In pulmonary cases where removal to the south was not feasible, the doctor would do well to prescribe a "flannel waistcoat and warm covering"[19] Where necessity required life in bad air, residents could "amend the Moisture of their Abode by removing Woods from them; they must also dry up their Ditches, keep their houses well aired with large Fires; they must exercise often and much; they must take less sleep than others, drink more liquids, to eat meals of warm nature." At other times, the journey might be replaced by "the warm bath, mild diaphoretics, especially diluents, warmer and more abundant clothing, particular flannel jackets worn next to the skin, frictions, frequent and moderate exercise."[20]

Such measures made artificial atmospheres commensurable with a change of air, demoting escapades to warm air that the pneumat Edward Kentish produced even in Bristol. Physician to the Bristol Dispensary during the 1810s, Kentish thought the French Revolution "a Moral and Political Epidemic," the cure for which he saw in a strict regimen and free ventilation. As the war stopped the English from going to the south of France, he sought public funds of 50,000 pounds to set up Madeira House in the south of England, which, with the adjoining grounds, was to be covered with an immense glass-frame where the air was to be maintained at a fixed temperature in which the "happy patient was to have been transported into a sort of fairy palace; without, indeed was a variable atmosphere, and all the miseries of the English climate; but within, the combined advantages of the steady climate of the south of France, and the genial climates of Naples or of Madeira." In Kentish's plan, the benefits of the air in Madeira House derived from correspondences between plants and animals; as plants were known to be "entirely dependent on climate," it was "probable equal benefit would accrue to animals by an artificial climate; it would secure those who come from a southern zone, and would impart genial influence of a more southern clime to the delicate and valetudinary of our own climate, who from delicacy of structure, may be regarded as exotics."[21]

Kentish's palace was a meteorological haven recreated in a town building. But "exotics" would not have necessarily liked it: Remove travel from cure and you might as well remove the whole idea of climate cure. To appreciate the complications involved in making sense of the logistics of air production and air change, we need to look at the arguments used to encourage, stifle, and otherwise influence the eighteenth- and nineteenth-century exodus of British patients to the Mediterranean in a quest for physical health, mental diversion, and rejuvenation of stamina. Scholarship on the Mediterranean, Middle East, Northern Africa, and the South Atlantic has enriched our knowledge of these spaces by taking them as cultural tropes in the real and imagined geographies fashioned by the past visitors, pilgrims, adventurers, salesmen, and local residents. My intention here is to offer a thick description of the means, methods, and words used to define these large entities as specifically therapeutic spaces and concerns over pathological exposure management during the early nineteenth century. Among other things, it can be shown that the ways in which disease etiology impinged on choice of resort and regimen meshed with an understanding of the region based not only on geographical and medical documents, but also on changing cultural stereotypes. It is therefore, in principle, possible to speak of the juxtaposition (perhaps even conflict) between the ostensibly scientific inquiries into the physiological effects of milder air and the economic, aesthetic, and patriotic underpinnings of British heliotropism.[22]

I begin by outlining the framework in which traveling for health acquired its nineteenth-century popularity and the ways in which it became associated with the Edenic and otherwise positive images of the Mediterranean. After examining how these images began to change under a barrage of criticism during the 1830s, I will identify the connection between such attacks and the budding disciplines of medical topography, climatotherapy, and research on consumption. While such connections may at first appear straightforward, conclusions about how this new knowledge affected medical rationales for health travel must remain tentative, partly because therapeutic advice was given in a nation that put a high premium on foreign travel and equable climates *regardless* of the doubts that the expert might have had about either. This uncertainty was enhanced by the fact that many resorts vying for patronage experienced shifts in fortunes as a result of changing medical opinion about their properties, while other, previously unknown spots, reached fashionable status and entered the big league. The stream of invalids and tourists meandered accordingly. Despite reservations, however, and regardless of the changes in professional views, one can still see a pattern whereby medicine and travel co-constructed each other. Charting the elements of such a pattern gives an important insight into nineteenth-century "alternative" medicine and "environmental" thinking during the period of ambitious public health works, industrialization, colonization, and racial anthropology.[23]

Mediterranean Promises

Eighteenth-century British health travel to the European and Atlantic Mediterranean was a well-documented phenomenon.[24] Yet the tradition was not new. Pliny and Galen wrote of therapeutic properties of land and sea voyages; consumptive Seneca went to Egypt in search of the Nile Valley's aridity; Cassiodorus, in the sixth century, preferred mountains.[25] By the early eighteenth century, the mobile habits of pulmonary patients spread in tandem with the popularization of sea bathing, hydrotherapy, dress reform, folk inhalants, and discussions on the possibility of acclimatization in the tropics. Colonial experiences had a momentous impact in the relativization of climate as a given and endurable by default; grasping the newly found zones of heat underlay the very logic of air choice, but also the xenophobic visions of the dark-skinned southerner, broiled to sloth.[26] Anxiety over miasmas and strategies of disease "avoidance" spurred costly measures to support the management of polluted space, as exemplified in the activities of continental "medical police."[27] In Britain, the "lure of the sea"[28] and the promotion of therapeutic effects of salt air, saltwater, and sea attracted wealthier clientele to Bath, Buxton, Scarborough, and Tunbridge Wells. Still wealthier clientele ventured to the European South,[29] in which case the travelers sought to combine therapy with tourism and pilgrimage.[30]

By the mid-nineteenth century, the very production of guidebooks on health travel was "one of the proofs that many opulent people in this country are bent on the endeavour to regain health, or to replace bodily uneasiness with more agreeable sensations, by change of climate, travelling, and mineral waters, instead of trusting exclusively to the ordinary medical treatment to which the majority are, with various success, most commonly submitted for the same purposes."[31] Change of air was considered as an (intended) misnomer referring to experiences as diverse as the alteration of scenery, diet, routine, and the mental immersion in classical cultures and foreign tongues. These took place in the ethnic realms of the Mediterranean people, whose social mores shook the choreographed etiquette of the English, even if they were "divested of the spirit-stirring feeling of the English."[32] The cult of the tanned body and the erotic appeal of anonymity in a dressed-down society began to emerge as powerful incentives. But while ordinary travelers might delight in this and other extravaganzas, the invalid was fettered by diagnosis. Where the former tramped a cathedral or joined a promenade, the latter kept strict hours and observed rules of exercise, regimen, and rest.[33] As nineteenth-century climatotherapists gained authority to specify destinations, such rigors set the tone of medical expatriation: "[T]he key to cure was heavy sacrifice."[34]

Was that sacrifice justified? Some British climatotherapists thought it was not. Central to developing this more critical attitude were medical exchanges about the merits of specific resorts in which the juxtaposition of pathological, meteorological, geological, sanitary, institutional, and cultural information produced a complex picture.

But such a picture met the needs of invalids with the widest range of afflictions, statuses, and motives. Mediterranean health tourism depended mainly on "young people with a family disposition to consumption; ladies fatigued by the London Season; persons suffering from chronic rheumatism whose pains and stiffness of limbs are invariably increased by cold; elderly people with bladder infections; other invalids undergoing wasting and gradually losing strength from disease of the kidneys; people with weak hearts [. . .], scrofulous children and persons recovering from some long debilitating illness."[35] How did climatotherapy create such a migration of bodies and metamorphosis of geographical perceptions that in some ways continue to structure the twenty-first-century European self?

One may first note a theoretical ambivalence: Eighteenth-century medics admitted ignorance about what constituted a healthy climate. More often than not, a southward migration was in reality an exercise in moral, social, and cultural reexamination. Contemporary medical topographies often charted the spaces of pain and health as if they were informed by the norms of transgression, embarrassment, and propriety.[36] Cold inspired violence, gaming, and alcoholism while it strengthened the body against putrefaction. Heat increased sensibility. When sympathetically transmitted to the mind, its overabundance defined the stereotypical southerner as passionate, amorous, vindictive, and indolent. Such correlations made it obvious that the fear of cold and love of warmth were insufficient to warrant unilateral travel prescription. But nascent uses of cold baths, exposures to "bracing" airs, and the doubts about the flannel "roasting" made it equally clear that one required more discrimination and caution in advising warmth therapy. If a sympathetic dermatology and respiratory thermometrics made sense among the professoriate of Edinburgh University, they still awaited a test in the real world.

Even by the 1820s most medical practitioners continued to wrestle with a correlation between physiology and climatological surroundings. Doctor James Clark believed that he lived in an age in which the notion that climate shaped health was only matched by a universal failure to explain why and how.[37] But Clark also knew that in dealing with meteorologies of health, the physician worked under the added burden of knowing neither the general causes of disease nor their local manifestation. He had learned this during a lucrative practice following his years in the navy and after a brief stay at the University of Edinburgh in 1817. In the following year, he chaperoned a troupe of Britons to France and Switzerland, where he gathered data on the effect of high altitudes on phthisis; subsequent practice in Rome from 1819 to 1826 allowed him to access high society, which led to the post of physician to Prince Leopold of Belgium and later to the Duchess of Kent. Esteemed and trusted at court, in 1837 Clark was made a baronet, but soon after the controversy over the case of Lady Flora Hastings (misidentifying an abdominal tumor as pregnancy), his practice declined. Nevertheless, Clark remained a household name and served on several medical commissions before retiring to Bagshot Park in 1860 in a residence let to him by the queen.

By this time, Clark boasted a strong publication record on climate and health. Launching his career with the Edinburgh dissertation "De Frigora Effectibus" in 1817, he produced a stream of well-received works in the next three decades, of which *The Influence of Climate in the Prevention and Cure of Chronic Diseases More Particularly of the Chest and Digestive Organs* merits attention for its programmatic statements on climatotherapeutic practice.[38] The book was redolent of an omnivorous Humboldtian approach in that it argued for a science of the organic and human worlds as shaped by "*all* the external agents which modify their actions." In Clark's opinion, even such an exhaustive approach would leave some matters unclear: While some things were amenable to calculation, many more were only fathomable by empathetic analogy. Confronting the deficit in knowledge by including *all* external agents might have looked foolhardy, but it was justified by Clark's insistence that the effects of the travel cure required knowledge of climatic *specificity* and a taxonomy of resorts based on morbidity statistics. If such statistics, by the blind law of averages, made Madeira as risky as Manchester, what sense did it make to put hopes in climatotherapy? When he argued this point in the 1830s, taxonomy of resort was only emerging independently of stereotypical adulation of the Grand Tour.[39]

This scarcity of information made a stark contrast with the popularity of southern exile that marked post-Napoleonic England. It corroborated the malcontents of the more conscientious medics that travel was an excuse for social visibility. Trite and sporadic at first, the objection stressed the faults of dissimulation and favoritism, which by degrees became the standard feature of climatological discourse on health. This mattered little for medical expats: Desperation, ennui, friendly encouragement, ready money, and prestige relegated any theoretical doubts. We know, for example, that James Johnson, an experienced traveler and a contemporary heavyweight in tropical medicine, expressed misgivings about the use and misuse of mobility therapy. He was determined to acknowledge lack of certainty in matters that had neither physiological nor statistical underpinnings and that showed no new ways forward. Guesswork ruled; fashions dictated; concerns about responsibility were negligible. Johnson sympathized with the fear of the French erstwhile military physician and jurisprudence expert Francois-Emanuel Fodéré: "ROUTINE has too much influence in the choice of climate," and neither patients nor doctors had a clear understanding of the factors on which their decision ought to be made. But although "FASHION and CUSTOM" ruled, it was cruel to deprive consumptives of the hope they harbored of prolonging their lives, as "this hope renders them happy during the whole of their journey to a foreign clime, and for a short time after their arrival at the place of destination. Such is the physical influence of this idea that they believe themselves cured—soon to be plunged in disappointment."[40]

The uncertainty allowed experimentation. Tellingly, Johnson called his 1823 fifty-day journey "an experiment the sole purpose of which was the pursuit of health and such amusement as was considered contributive of that object."[41] Such experiments established travel as a form of antidote to modern versions the "English malady" that Johnson dubbed as "wear-and-tear." He described it as an asthenic condition "intermediate between that of sickness and health, but much nearer the former than the latter."[42] It resulted from social metabolisms nurtured on the sofa and in the salon, linked to hypochondriasis and sedentariness; it grew out of night watching and improper diet, and was created by overwork and anxiety. Its worst signs showed up among female youths whose conviviality, gossip, and deep-cut toilette smacked of long evenings and recirculated cameral atmospheres. According to Johnson, this specifically *female* "phoebophobia" (fear of sunlight)—pandering to the cult of aristocratic paleness—bred the "lifeless Albinos of the Boudoir," whose concerns with premature wrinkles in sooty London air prevailed over outdoor amusements. It was evasion of "natural" life that the Mediterranean destroyed; it was the solution for "the epicure, the gourmand, the mere man of pleasure, the overstimulated over-cultivated woman, the anxious care-worn professional man, the toiling, striving, struggling man of business."[43]

Choice of locality was crucial. Every ailment required a corresponding climate, regimen, and season. Slow digestion, hypochondriasis, and disposition to hemorrhage generally improved in Naples and Nice. Dyspeptic patients needed Italy, while Rome helped those with gastritis. Pisa was recommended to individuals with intolerance to weather and wind. Wounds healed faster in the Nile Valley. More generally, the effect of almost *any* sunny landscape seemed to arrest the progress of consumption, among the most devastating chronic condition in Britain. Before the middle decades of the century, medical opinion described the disease as inflammatory. Its sufferers were advised to live in a sedative environment of mild and dry air. Winter residences included Madeira, Azores, Tenerife, and Tangier. These were not just salubrious spots; they were also exempt from the turmoil of civilization. During the Franco-Prussian war in 1870, medical correspondents assured the reluctant public that Cannes, Nice, and Mentone remained calm, peaceable, and open to consumptive patients, because the resorts' very existences depended on the winter influx of invalids and visitors. A stress on equability of weather and civic peace became a key element in health-guide writing.[44]

Contrasting the healthy South with the albinos' Albion spurred a culture of hyperbolic approbation. It colored travelogues and nurtured wild expectations.[45] However incoherent or unrepresentative, hyperbole spawned a belief in the supreme healing prowess of the Mediterranean coasts. Veneration was considered pathetic. Mary Shelley rhapsodized about an evening experienced on a mule's back above Amalfi when Nature imparted "a quick and living enjoyment akin to the transports of love and ecstasy of music—it touched a chord whose vibration is happiness."[46] Humphry Davy loved "the mild climate of Nice, Naples or Sicily, where even in winter, it is possible to enjoy the warmth of the sunshine in the open air beneath palm trees, or amidst the evergreen groves of orange trees, covered with odorous fruit and sweet scented leaves, *mere existence is a pleasure*, and even the pains of diseases are sometimes forgotten amidst the balmy influence of Nature."[47] Several decades later, and more radically, Charles Barham, president of the Royal Institution of Cornwall, put forward the idea that the correction of constitutional disease *required* climatic agency. This made a removal to a foreign locality a medical priority, because "nothing to be had, naturally or artificially, in this country, can with equal advantage be substituted for it."[48]

It should be remembered, however, that such representations derived more from a "poetics of pathology"—an aesthetic-cum-medical description that was enshrined in the Baedekers of the time—than it did from medical maps or morbidity statistics, if there were any to be had.[49] Italy and Greece held a special place in romanticized geographies of British valetudinarians. For them, such loci "connoted magnified possibility, mystic revelation, and transfigured destiny."[50] Rome, the magnet city, epitomized these possibilities, attracting seekers of both sun and catacombs. "The visitors to Italy and the Eternal City," reflected *Bentley's Miscellany*, "are now to be found wending their way—some to realize on the very spots all the dreams of study—the artist to gaze on the glorious monuments of art [...] the *fast* man to *do* Italy in the shortest possible time, and with superficial mind to gaze and forget."[51] The praise of Rome's air meshed with that of the city perceived as the hub of classic art, political virtue, and Roman Catholicism. In fact, it was even contended that Rome *owed* its achievements to quality of air, which, despite its ubiquitous stench and grime, continued to have a preternatural effect on pilgrims and invalids alike. A diarist compared its atmosphere with a "sweet cowslip wine,"[52] while in her travelogues, Hester Lynch Piozzi remembered a gentleman who, commenting on "Padua la dotta [as a] very stinking nasty town," held nevertheless "that literature and dirt had long been intimately acquainted and that this city was commonly called among Italians Porcil de Paudai, Padua the Pig Stye." For her part, Piozzi found it logical that the Paduan asses had a prodigious milking capacity, as they were bred in an air "where every herb and every shrub breathes fragrance." British consumptives should not shrink from this nectary latte as it "liquefied" the air into a potable medicine.[53]

Hazards Unmasked

Increasingly, and despite the mass exodus propelled by cheaper transport after the end of hostilities, even the most enthusiastic of practitioners warned against the perils of journey, food, poor sanitation, and the excitement occasioned by change of scenery and society.[54] Cautious practitioners considered it imperative not to overexert the patient, especially the female, who, as "weaker" and less accustomed to the hazards of exposure, stood slimmer chances of surviving the odyssey. The timing of departure was crucial, the diagnosis decisive. Benefits would be reaped only if travel commenced during the acute phase. As the condition preceding a disease could pass unnoticed, a consumptive patient could develop a cough and blood-spitting during the journey. To prepare oneself before departure was critical, and the advice was to calm the digestive tract by moderate drinking, riding, warm baths, and cold sponging. Milder climates could arrest the progress of disease only in sync with a comprehensive prophylaxis.[55]

Another concern was the Mediterranean itself. Information had already become available on its darker sanitary side, its proximity to deadly African winds, and the unaccountably high humidity and rain. In one of the most extensive literary commentaries on the patient's progress through the Mediterranean, Tobias Smollett found it hard to accept that a city as pleasantly located as Nice would experience fifty-six rainy days during the four winter months of 1767, "a greater quantity than generally falls during the six worst months of the year in the county of Middlesex." The approach of summer brought new risks, because heat rarefied humors and opened pores to experience the northeastern winds that had sucked in the cold while passing over the late snows of the Alps and the Apennines. "Even the people of the country, who enjoy good health, are afraid of exposing themselves to the air of this season," and, if asked to spend his life near Nice, Smollett would choose a hilly station where the more fitting cooler air would guarantee freedom from the air's salinity during summer months, and where the air is "unmolested by those flies, gnats, and other vermin, which render lower parts almost uninhabitable."[56] Naturalist Patrick Brydone, an early Scottish traveler to Sicily and Malta whose work met with critical acclaim among contemporary vulcanologists and electricians, observed that the Sicilian climate was "by no means what we expected. And the serene sky of Italy, so much boasted of by our travelled gentlemen, does not altogether deserve the great eulogiums bestowed upon it [...] I am persuaded that our physicians are under some mistake with regard to this climate."[57]

Within half a century, a litany of warnings filtered into the public to mold an opinion about the discomforts, dangers, and distempers lurking along the southern European shores. Medical travelers were struck by the incidence of local fevers, marshy effluvia, and the unsanitary microclimates of resorts.[58] Regarding the air of Rome, a sardonic reviewer for the *Edinburgh and Medical Journal* remarked that its quality had nothing to do with the atmospheric reality. The Holy City, rather, possessed the best climate in the Occident because it was placed under the care of the Holy Virgin and any attempt to challenge this assumption "would have been regarded as something profane and blasphemous, and would have been rather dangerous for the person who attempted to do so." Had it been the case that Rome precipitated disease, as it was believed, "the simple rumour might have diminished the revenues of the Pope and Cardinals, and all their dependants, to an extreme alarm to contemplate, and quite ruinous in its effect."[59]

In short, no respectable early nineteenth-century physician would pass a definitive judgment on what constituted a good climatological resort. In 1806, Beddoes penned a pointed criticism of how climate doctors manipulated such agnosticism. His *Manual of Health* gave a bashing to the purveyors of "airing business," whom Beddoes accused of unashamedly using climate as a smokescreen for ignorance. What resort therapy did best was to send desperate cases away from sight into decay, into charnel houses: a practice "undoubtedly very convenient for the baffled practitioner," but good also as a means for fulfilling another purpose of medical art, diversion of the sick. "For what, I ask is this art? [if Hoffmann and William Cullen defined] medicine [as] the art of curing and preventing diseases, it is still more the art of amusing those patients, whose diseases it can neither remove nor palliate. Those who amuse the patient have more success than those who cure the patient; medical attendants frequently aim more at amusing than curing."[60] For Beddoes, this was a predictable plot: A physician attends a consumptive person and, on seeing his disease in advanced stages, keeps his hopes up with placebo tricks and diversion regimens. These, by degrees, coming to an end, the doctor faces the moment of truth, but feeling that it would be ruthless to leave a dying person without succor, he recommends a move to "Dawlish, Exmouth, the Lands End, or God knows how much further of still." The family doctor must find this strategy convenient in having a recourse to a "charnel house" into which to "toss a half-animated carcass" for the care of his fellow physician at other people's expense. If a patient expires in a resort, the physician could always explain: "I advised him to go so and so, but they could do nothing for him."[61]

While the culprit was often a capricious patient, the ultimate villain was nature. Physicians were careful to warn that, notwithstanding deceptive averages, a southern European destination nurtured both scorching summers and noxious winds. The latter were particularly abhorred. Thought to originate in the sands of Libya, Arabia, or Egypt, winds with bizarre names like the sirocco, *leste*, *shamal*, and *levant* (*jugo* in the Adriatic) infested Europe with suffocating, depressing, and "relaxing" qualities for which the civilized constitutions had no defense. Sirocco in Naples stopped digestion, killed gourmands, and drained plethoric constitutions; Palermans shut houses and stayed inside for fear of being assaulted by erysipelatous inflammations. During one sirocco onslaught, Queen Caroline of Naples, according to her English confidante, had to crawl from a marble floor to write a note for a servant, and then "throw herself down again, in order to alleviate the oppression she felt: such was the inconvenience endured even by royalty [...] in the otherwise enchanting valley of the Conca d'Oro."[62] Change of winds exhausted the constitution and caused unexpected symptoms of disorientation, as witnessed by Richard Cobden, who, after a violent storm at Desenzano, remarked: "[T]he effect [of the storm] was instantaneous, and excepting the absence of motion our sensations were much more as we had experienced a gale of wind at sea on board of a vessel. I almost felt sea-sick when I went to bed."[63] Such fears fueled the notion of an intra-tropical poison belt that churned the winds, "involving all things in a suffocating heat, sometimes mixing all the elements [...] and sometimes destroying all things in their passage." In temperate climates, physicians tended to take it for granted that European epidemics had been "imported from some sultry climate."[64]

Climatotherapeutic adventure was on occasion perilous in outcome. Not only could the journey itself prove tantalizing, but also advanced stages of disease, weariness, and vagaries of new air in the places where such vagaries were not expected could turn fatal in a matter of days. "I begin to suspect," was the inspired insight in Henry Matthew's *Diary of an Invalid*, that "all I shall gain by my voyage will be the conviction that a man who travels so far from home in pursuit of health travels on a fool's errand. The crosses he must meet on his road will do him more injury than he can hope to compensate by any change of climate."[65] Topographers confirmed such warnings: "I have long lived and seen too much, not to know the errors of discrimination and the fallacies of hope that send pulmonary invalids from the gloomy skies but comfortable adobes of England to lands where comfort is unknown, even by name, and who atmosphere cannot work miracles, whatever their saints may do."[66] Such misgivings increasingly plagued nineteenth-century opinion: With every coffin dispatch, the medical South took on a grimmer aspect. Disappointments found in travelers' letters, diaries, and official reports meshed and amplified with the crankiness of the homesick valetudinarian.

At first something of a surprise, the repeated complaint showed in gory detail how the Mediterranean weather could quickly overmatch the British in inclemency. Pessimism followed in the steps of findings such as those that found Lombardy "deformed by the severity of its winter"; that Rome's January temperature regularly fell below zero, and that Naples's spring could be described as "extremely rigorous." Provence and Nice stayed miserably cold until May.[67] Pisa and Cornwall had equal rainfalls, but Sidmouth had a milder January than Campania, where the thick-walled vernacular architecture served the purpose of protecting from heat rather than cold. Extremes in either direction were too trying not to be mentioned: Johnson's fellow lady-traveler captured this in a morose snapshot of "Rome and its surrounding deserts, [where] every thing depicts the death of Nature; in Naples, and its environs [...] a feverish vitality [...] consumes while it brightens. The air is fire, the soil a furnace.

Sun-beams bring death!"[68] While this might have been uttered by an unreconstructed "albino of the boudoir," it resonated for disillusioned English consumptives who, like Henry Matthews, felt that Naples had "one of the worst climates in Europe," which, for the absolutely hopeless cases, could be used as a method of voluntary euthanasia.[69] As in the discussion of other low-lying regions, miasmas were the worst threat.[70]

From around the late 1820s, however, another, more systematic criticism, began to spoil southern ideals because of what its champions presented as a statistical and scientific cross-examination of evidence related to weather and morbidity for a select number of resorts. New medical researchers began to seek attention of the public and the profession by using whatever quantitative observations they could put their hands on to stifle the hearsay, which, in their view, had spurred a facile veneration of southern Europe among British expatriates. For these statisticians, only the actual meteorological conditions might dispel (or confirm) the vernacular perceptions and determine which, if any, factors possessed curative powers sufficient to justify the popularity of the change of air. Did these have any effect on, for example, consumption? Were they to be found *at all* in Nice, Rome, Madeira, San Remo, or Tangier? How could their effects be compared with other, nonquantifiable unknowns such as sensual stimulation and scenic variety? The time of this reappraisal, it is worth remembering, was exactly coincidental (and in part caused) by the investigations on consumptive recovery at high altitudes and colder air, which, from the 1830s, brought increasing attention to the virtues of rarefied atmosphere in alleviating pulmonary conditions.[71]

The first flurry of critical statistics appeared in the wake of reports on the health of the navy during the 1830s that contained morbidity and weather data. The latter included temperature, humidity, insolation, winds, electricity, and air pressure. While the results for the most part corroborated the standard view, they were often at odds with it. Comparing morbidity statistics with patterns in Gibraltar weather, for example, a surgeon alerted his readers to the fact that "the effects of the climate in the prevention or remedying of some diseases, particularly those affecting the lungs, are perhaps not so great as they have often been represented."[72] Accounts by military and civilian doctors such as Major Tulloch, John Hennen, and Henry Marshall gave an up-to-date assessment of the role played by consumption. Their conclusions challenged the risk attributed to changes in weather, indicating lower mortality for the more variable Ionian Islands than Malta and western Mediterranean.[73]

More important still were changing views on the etiology of consumption. James Clark understood consumption to be inflammatory and as such in need of sedative treatment in relaxing resorts like Pisa, Naples, Rome, and Malaga. But by the 1830s, medical theory moved on to consider debility and malnutrition as the culprits: Relaxing a consumptive even further in a warm, calm, and muggy Campania was more likely to obstruct than help recovery. In the otherwise sympathetic review of Clark's *Influence of Climate*, an unsigned reviewer for the *Times* questioned Clark's omission to note the "opinion which is gaining ground" that dampness rather than aridity could prove of main benefit to a consumptive traveler. As in other arguments involving considerations of geographic nature, information on endemicity (or lack thereof) ruled supreme—rarity of phthisis in fenny regions seemed sufficient to warrant the exclusion of moisture from a list of exciting causes. By midcentury, the work of physicians and medical meteorologists such as Alfred Haviland, Thomas Burgess, William Scoresby-Jackson, and Henry James Bennet provided a fuller synopsis of the geographical distribution of consumption, now only weakly correlated with cold, moisture, and variability of the weather. By this time, it had also become mandatory to stress the importance of outdoor exercise, not the passive external condition, as the mainstay of recovery. The change in habits and exposure to air mattered more than a direct physiological reaction to physical circumstances.[74]

Commenting in retrospect, James Alexander Lindsay, physician to the Consumptive Hospital at Belfast, noted the odd fact that Norway, Holland, and Italy had identical rates of mortality from pulmonary disease. There was also the fact that the wettest spots in the North Atlantic, such as the Hebrides, Shetlands, Faroe Islands, and Iceland, were virtually free from the condition; breathing moist air, it seemed, had no role in predisposing to the disease. Neither did sharp weather, as shown in the low mortality figures in the Alps and the Scottish Highlands.[75] Following on the heels of this counterintuition, medical advice swung toward the more "bracing" and "tonic" climates characterizing the drier, windier, and cooler resorts such as Montpellier, Nice, Menton, and San Remo. But this shift also gave a boost to British and other North European resorts such as Norway. Lindsay wrote: "[T]he popularity of the marine resorts, once unequalled, and still great, seems slightly receding in favour of the modern drift of medical opinion in favour of the view that they afford on the whole less favourable results in consumption than the mountain sanatoria, the dry inland resorts and the sea-voyage."[76]

Portable Climates

While nineteenth-century mass culture favored the South indepen-dently of experts' objections, the medical traveler needed assurance that her funds were not to be spent on her own death. Science's ostensible impartiality was best for this purpose. But how exactly did scientific medicine go about fathoming the curative powers of the Mediterranean world? We begin with London practitioner John Charles Atkinson, one of the young Turks who made good use of newly available navy reports. Atkinson was a member of the Royal College of Surgeons with long-standing interests in epidemiology and public health. His works during the middle decades of the cen-tury explored the correlation between cholera and atmospheric elec-tricity, and the therapeutic effects of wind, sleep, and fear.[77] During this research, Atkinson offered a caustic view about the scope and efficacy of air cure, especially with regard to the bracing French Riviera. He called climatotherapy a fad, indicting physicians not as much for trying to evade the charge of incompetence, as Beddoes had accused them, but for making the change of air a panacea that vied with other exploded remedies such as elixirs, drops, waters, and homeopathy. "Change of air" was forced upon everyone: Public and professionals advised spas, localities with healing powers, and expo-sure to fresh air, even though common sense counseled that in most cases of confirmed phthisis, it was the "antagonist" action to the lung process that was needed, not fresh air! How come, to take one glaring example of inconsistency, the air of Rome got such a good press when the whole of Campagna exuded malaria? Atkinson's rebuttals of the method were especially acerbic in relation to what he perceived to be faulty advice: Those putting trust in a change of air were giving credit to a placebo and scandalously disregarding navy reports that showed higher morbidity from consumption in warmer regions than those in the midlatitudes.

Neither the sun nor the outdoors was enough in itself. Atkinson's experience confirmed this. He prescribed home therapy to a girl patient, who ignored it under the pressure of her friends. She was transported to southern France, where her fever returned in three days; two days later she expired. Atkinson believed that the still unrecognized virtues of indoor climates were preferable to the effort that would be required to relocate a patient to a resort. As Beddoes had already advocated, airiness and equability could be found in the home and were superior to expatriation. The indoors mimicked the Mediterranean physical qualities, but the real benefit accrued from an easier observance of prophylactic measures such as clothing, diet, hygiene, heating, and ventilation. Heating and ventilation were already available for domestic use. Others knew that preventions against cold, catarrhs, and consumption lay in warm clothes and warm living quarters.[78] Prevention could even rely on sheer habit. Challenging the ultra-sanitarians, Atkinson presented the case of a consumptive who, having spent his whole life above a Thames sewer, gave in to his physician's pleas and changed residence to a less grim spot. This worsened his condition: "[O]ur forefathers were right to suggest that consumptives should visit stables, farm-yards, sheep folds, cow-sheds early in the morning because the carbonic acid there would alleviate their condition; the Statistical Society has in fact proved that scavengers and nightmen have better health and appetite than other workers."[79]

For those whose condition permitted more active life, newly developed mechanical respirators came in handy.[80] The leading innovator here was Julius Jeffreys, a Humboldtian generalist with interests in "pneumatic physiology" and an inclination toward medical astro-meteorology.[81] Jeffreys witnessed the effects of tropical heat on the "irritable state" of the lungs. His interest in the effects of the atmosphere began with a trip to Himalayas, after which he made a recommendation to the government to establish a high-altitude sanatorium. But he also considered the solution to exposure in a more practicable light. Medical practitioners, he proposed, should supply lung sufferers "with a local *artificial climate* suitably different from the general atmosphere around his bodily surface, which is not irritable."[82] This led to the respirator that Julius initially modeled on the Royal Humane Society's humidifier for asphyxia and that he improved with the view of helping his phthisical sister, Harriet. The respirator, which was attached to the face, was a cone-like contraption supplied with iron wires designed to absorb heat from the breath and act as a buffer between the lungs and external cold. The device reduced climatotherapy to a science concerned with the quality of inhaled air and the "constant exposure" to "fomenting climate" regardless of where and by which means such warmth could be produced. The body might have been in London, but with the respirator the lungs were in Malaga.

Jeffreys preferred this decoupling over relocation because the body was only marginally affected by warmth; for pulmonary invalids, in fact, "no natural climate is exactly suitable." They were better suited to an inexpensive, "portable climate": "[T]hese respirators are small in bulk and weight and that renders them perfectly portable, and therefore the climate they afford is portable." More generally, it helped with intolerance to sooty fogs, prevented sore throat and painful expectoration, and alleviated hoarseness of voice and tickling in the larynx.[83] Jeffreys's marketing of the invention relied on its convenience and size—the *Times* listed it in an ad for "Auxiliaries of Comfort"[84]—even if worked against its weird street appearance. In this he was going directly against the surge in foreign health travel, as is clear from the users' letters of praise that noted the respirators "ability to entirely replace residence abroad." Even in Pau, Nice, Hastings, and Torquay, promenaders found it indispensable during harsher weather. Jeffreys wondered if this preference could be a sign that there really wasn't a natural climate that could match that of the respirator.[85]

Needless to say, such gadgets must have looked like poor surrogates for talismanic resorts. Atkinson disagreed, pushing for a method that was even more radical then sporting a conical beak on the Strand. Using data on the influenza of 1847, he discovered that more people had been affected in cities than in suburbia. He also assaulted the conventional "fresh air" wisdom and denied the agency of pure air in both the etiology and therapy of bronchial diseases. He thought the outcry against carbon dioxide ludicrous: Eskimos, Russians, and other northerners never suffered in their tightly isolated rooms. Why should Englishmen? His own practice showed that people not only found advantage in being confined in airtight rooms, but also had their health improve.[86] Atkinson had in mind the Irish famine fever of 1847, when more poor died in the hospital wards than those left exposed (almost naked) to the elements, because, in his view, those who from birth had enjoyed the open air could later suffer from extended exposure to the indoors. Nature seemed in this instance to "fortify itself against external agencies," or, as Johnson remarked, the "perpetual scene of atmospheric vicissitudes not only steals us against their effects, but proves an unceasing stimulus to activity of body and mind, and consequently, to vigour of constitution."[87] Therapy thus needed to take into account national, perhaps racial, differences: English, well fed and well clad, habituated to comfort and indoor activity, could not cope as well in rain as the Irish.[88]

Such arguments made sense of eudiometry's failure to discriminate "between the purest air of mountains and that from the pestilent courts of a crowded city."[89] To forcibly "eject" a patient from the soothing air of the apartment in the name of experiencing the unsullied air of the Mediterranean was to commit a fallacy based on travelers' tales. Such advice neglected a spatial frame of disease, as "our functions of life acquire habits, so to speak, according to the

position of the individual in the world."[90] Here it was not only location that determined endemic diseases, but also the habits and the
customary use of nonnaturals: A snack for a Dorset farmer would be
a feast for a London beaux. An indoor life could only with risk be
converted into one lived outdoors. Disposition to disease depended
on long-term exposure to native weather and soil. Change of exposure would jeopardize the constitution. Atkinson even admonished
practitioners to abstain from condemning fog and humidity as long
as they demonstrated even the slightest benefits for longevity. As the
early rising and breathing of morning mists had always been a rule
among the very old people, Atkinson wrote, their practice proved the
healthy effects of damp atmospheres.[91]

The Madeira Debates

From midcentury onward, the critical approach to medical travel continued to sharpen its criteria. Four years after Atkinson's work, in
1850, Thomas Burgess, physician to Blenheim Street Dispensary in
London, attempted further deconstruction of therapeutic change of
air in the unambiguously titled "Comments on the Inutility
of Resorting to the Italian Climate for the Cure of Pulmonary
Consumption." Burgess was an acne expert with interests in the sympathetic theory of blushing and the role of mental activity in capillary
motion.[92] He built his mature work on the experiences at the dispensary where he first realized the extraordinary magnitude of interest
in and popular misconceptions about Mediterranean climatotherapy.
What pained him was not its popularity with the rich, but the belief
of his poorer patients that the climate cure remained far beyond their
reach: "I am told that I would get well if I could only go for a short
time to some warm country." He viewed this hope as a chimera. It
was shared by those who still believed that consumptive "patients
were sent to breath the volcanic air, drink milk and die."[93] There
was sadness in the naïveté of *pulmonaires* set to spend weeks on a
journey to the climatic "bubble," for no other purpose than ruin the
lungs by the brutal change of surroundings. It was the change, not
the air, that made "change of air" such a dubious business. Instead,
for most sufferers it made more sense to stay at home and, if such
was their fate, die in the bosom of their families rather than populate
Italian graveyards.[94] As Johnson indicated, the illusion was fostered
by a blind parental care,[95] or, according to a *Lancet* correspondent,
in the "innate roving disposition so characteristic of our countrymen as to have become a feature in the national physiognomy."

More disquieting still was the professional negligence born out of "the ennui occasioned by protracted attendance upon a ruthless and immedicable disorder."[96] Climatotherapy, in Burgess's view, lived off the rambling patient and the doctor.

Burgess's real target was not malarial Rome but Malta and Madeira, the epitomes of purity, health, and equability. His purpose here was to dispel the then current opinion about these properties, and to show why almost any change of air might be potentially unsafe. At stake was no less than the basis of pulmonary climatotherapy. In the Maltese case, Burgess used statistics from the Valetta hospital to show that consumption deaths there rose three times above those from the second killer, fever. Of 108 cases of respiratory disease, the majority occurred in the warm and presumably "benign" months of May, June, and October, and thirty three percent of soldiers died from palsies in one year, which was just about the average. Furthermore, meteorological averages should be regarded as having a minor role in the evaluation of a resort; rather, it was extremes that mattered. In other words, despite Britain's chilly climate, it was still benign enough to spare consumptives from foreign airs. Not to mention that the Maltese winds were known to deplete vitality and lead to copious dews. "I do not remember ever to have felt the sensation of cold so acutely in this country (England) as I have done in Malta during a dry, North-westerly, or north-easterly wind ('gregale')."[97]

Burgess's main interest, however, was to give a medical account of Madeira, "the Island of the Blessed." This entailed a dissection of climatotherapy as a discipline. In questioning Madeira's presumed virtues, Burgess first undermined the received view on the benefits of thermal uniformity and dry air. Early weather reports from the island, like that produced by Joseph Adams (1801), made it a reputable destination in which local residents boasted an average age of thirty-nine, as opposed to nineteen in London.[98] Boat routes linked it to Liverpool in eight days and helped bring it within reach of those who tried to "cheat the winter of its own climate." American consumptives were also regular visitors. The island seemed to have an impeccable meteorological record: "Hence a drop of dew seldom falls, except in the higher parts of the island, and any deleterious effluvia, which may arise from the surface of the earth, or from other sources, are dissipated as soon as they are produced."[99] Its thermal uniformity was its best-selling feature, especially among champions of the inflammatory theory of tuberculosis, which kept the patients "wrapped up in flannel in rooms maintained at a Madeira temperature."[100] On this basis, authors like Johann Heineken and James Copland preferred it as a phthisical resort over any other place of southern Europe: A "Madeira-House" for treating lung disease was even established at Bristol.[101]

Despite this association, however, the real Madeira failed to become a mecca of climatotherapy. Lack of accommodation and institutional support, a monastic social life, and the distance from continental Europe robbed it of a wider popularity. It was out of the way. Nor was its weather particularly good: "[T]he winds enter it almost without interruption, to the great dread of the mariner, [...] bad and a continual surf beats the beach."[102] "A sultry mist hangs heavily, The water, air, and sky, Wear each the same dull, sober gleam."[103] However, medical experts undermined the island's reputation, which led to a controversy that erupted in the 1850s. Seemingly about the beneficial qualities of its climate, the controversy widened to include the proper place of science in medical travel and professional credentials. To see exactly what was at stake, we need to outline of the arguments of three main protagonists, John Abraham Mason, Thomas Burgess, and James Mackenzie Bloxam.

The landmark publication was a study of the island undertaken by John Abraham Mason. Mason studied in Paris and earned a medical degree at Edinburgh. Diagnosed with tuberculosis in 1820, he turned to James Clark. Clark recommended Nice. Mason, his wife, and a relative set off for France, but on arrival in Dieppe, the relative fell ill with brain fever. The party stopped for six weeks, just long enough to miss the season. Mason consulted Clark, who suggested Madeira. The Masons moved on promptly without returning to England. There they stayed for two years, during which Mason suffered from boredom as much as disease. For a consumptive, however, he immediately launched an ambitious, if ultimately self-destructive program of observations on every conceivable medico-topographical feature of the place. He even invented a hygrometer. He observed geology, hydrology, and natural history, tested medical springs, took readings of temperature, humidity, wind, and insolation at least three times a day, sometimes as many as twelve. He kept a medico-meteorological journal and calculated averages, planning to put them together into a definitive climato-medical survey of his adopted home.[104]

Two years of research produced a stunning quantity of information. But for all his enthusiasm, Mason's health barely improved enough to allow him to move to the more bracing Nice. The couple went to Le Havre by sea, and then by land to Avigno. There they took a diligence to Nice. The last leg of the journey was rushed and ill planned: Fruit and stale bread were the staple diet. The weather turned sultry, water became sparse, and the roads grueling. Arriving in Nice, Mason was stricken by dysentery, which, two weeks later, claimed his life. He was twenty-seven. On hearing the news, Clark asked for an autopsy. The findings identified dysentery rather than consumption as cause of death: The lungs, the report said, could have survived for another five years.[105]

Mason's views on climatotherapy embraced contemporary physiology. His posthumous notes showed an inordinate amount of analysis on the role of atmosphere on animal economy based on the notion of sympathy between external circumstances (milieu) and operation of an organism. Mason worked within the context of physiological checks and balances. When healthy, the human constitution remained immune to the influence of meteorological conditions. But when the "correspondence between the animal organization and the accidents of a range of climate" broke down, the sufferer was entitled to seek its restoration in a different place. The key criterion in finding such a place for pulmonary cases was its "hygrometricity," that is, seasonal changes in the absolute and relative amount of water vapor. Increased perspiration was crucial in the cure of consumption. Mason's experiments were designed to determine the comparative hygrometry of London and Madeira and decide whether the latter qualified as a legitimate retreat for consumptives.

Among the more controversial of his findings was that the island rarely benefited consumptives. It was only *slightly* drier than London (during June and July 1848, it was actually wetter). Unlike London, it suffered from moisture-laden African winds, which, apart from miasmatic qualities, eroded table utensils and furniture. According to Mason, it was a rare occurrence in the winter months to see a sky clear, while the properties of the *leste* kept people indoors, dispirited. Mornings were cold, dews plentiful. Yet while such findings seemed damning, they were not brought forward to "prejudice people against Madeira, but to alert them on overreliance to sanatory effects of its climate." It was this overreliance that was wrong, not the humidity. Wording his distinction carefully, Mason claimed that successful treatment would only consolidate Madeira's reputation, but "the failure of cases, to which its climate is not adapted, would not be attended with the effect of damaging its character, as a residence for those who, by a change to such a locality, might reasonably calculate upon the realization of their most sanguine expectations."[106] The opposite was often true. Both success and failure could be damaging to resorts: Success might create an unreasonable expectation and failure might suggest climatological risk.[107]

The reception of Mason's work, edited and published posthumously by James Sheridan Knowles, the playwright and medical practitioner at Cork, almost entirely glossed over its subtleties and took it as an outright denunciation of Madeira. Burgess's use of Mason was a vehement assault on the latter's interpretation of warm climates. In a series of articles for the *Lancet* published in 1850, Burgess pulled together all available evidence to deny the "utility" of medical change of air. He thought that this could be borne out by statistics, observations, and case histories. It could even be substantiated by a theological perfect fit, as "nature has adapted the constitution of man to the climate of his ancestors"—to his "hereditary climate"—and that the most reasonable course of action for a patient would be to undertake a "change of air in *his own* climate," or removal to one nearly approaching to it.[108] Burgess asked if it were not consistent with "Nature's laws" that an English consumptive should be cured in his or her own climate. Why should a warm climate be preferred to a cold one, if there were no variability in either? And how many people would need to die during the "probationary process" of travel to prove this point?

The body had a memory of place. Once removed from that place, it would be haunted by a pathogenic nostalgia.[109] Burgess proposed a more drastic understanding of the "sympathetic" interconnectedness of local peoples and natures. Whether to credit providence or heredity, nineteenth-century medical practitioners had little doubt as to the reality of correspondence between place and local nature, whether in plants, animals, or diseases. It was, then, a matter of ingenuity (or disingenuousness) to find a reason for foreign health travel without facing the charge of violating the natural fit. Mason's hurried providential reflections testified to the relevance he bestowed on his experiential pneumatics; the influences that surroundings have on the animal economy, he maintained, ought to be considered primarily with reference to healthy organization and harmonious exchange. In health, the atmosphere had little power to alter either the subjective or objective state of the body, not withstanding the negative effects of adaptation. It was only in ill health that air could harm, when there no longer existed "that correspondence between the animal organization, and the accidents of a range of climate, which is necessary to ensure a due performance of the animal functions. In such a case we are bound to take advantage of a peculiar locality."[110]

As for Madeira, another of his targets, Burgess refused to give grounds even to the proposition that the island's climate could benefit patients with advanced phthisis and for whom colder weather would be detrimental.[111] He rebuffed Mason's detractor Robert White, "a non-medical writer," who in his reports on Madeira had accused Mason of mishandling instruments and yielding to the depressive moods occasioned by advanced pulmonary condition.[112] Where White protested against Mason's statement that "under the morbid influence of active disease [which made him complain] bitterly of the cloudy sky, the high winds, and the variability of the climate of Madeira," Burgess replied, on the contrary, that White was himself deluded—"evidently grateful for the blessings of his restored health."[113] For Burgess, Mason's conclusions commanded respect as they "were the subject thoroughly investigated, few places would be found where the system is more liable to general disorder; while, at the same time, I suspect that the average duration of life would turn out to be inferior to that of our own country."[114]

Burgess's reading of Mason's empiricism and his qualified criticism of Madeira raised a brow or two. It disgusted James Mackenzie Bloxam, a Denbighshire barrister and a regular visitor to the island. In vitriolic response to Burgess, he presented himself as a patient, not a medical man of science, because the misuse of instrumental readings in climatotherapeutic literature had "created in me a distaste for pursuit of science."[115] The thrust of argument was to challenge all science in its professed ability to detect the influence of climate on disease, especially when there were other, independent methods, whereby these influences could be demonstrated. Most people, he wrote, seemed to think "that a few hygrometric and thermometric data are sufficient to enable them to pronounce that climate of long-established reputation, whether good or bad, is in fact the reverse of what was previously supposed." Data, however, could be made by any instrument: Badly calibrated instruments could be read by people without qualification and experience. Finding a proper place for the hygrometer was no arbitrary matter, nor subject to accident, proximity, or convenience. Neither was any particular spot chosen as representative of the island's climate—there was no way to locate Madeira's "average weather." There was also the fact that the epithets describing the weather—especially in terms of its dampness and dryness—were rarely less relevant than instrumental readings: "Is it not obvious that the sensation of dampness depended on something besides the number of grains of water, in a cubic foot of air, and that dr Mason's satisfactory manner of estimating dampness fails to detect that not unimportant something whatever that may be?"[116]

Bloxam didn't like Mason's editor's idea that patients would benefit from keeping a thermometric record. What was the point of this practice? Why nurture an "exclusive faith" in meteorological information? If anything, it would mislead the physician if unaccompanied by a clinical picture or by a patient's own reports. It made little sense to use science in cases where common knowledge and feeling were sufficient. It was also pointless to seek the help of meteorologists in situations in which patients might be more qualified to pass a climatological verdict. It was equally wrong to accuse local physicians of not being scientific enough. This was no different from charging a man with not looking at this barometer on a fine day under a dread that its index will point to "foul weather." Tongue in cheek, Bloxam suggested that London medical men should really prefer an instrument that would, like the barometer, indicate " 'consumption, fever, cholera' at different parts of hygrometric scale and to prescribe climate for their patients accordingly?"[117] Why should anyone accept these "cabalistic numbers" as ultimate authority? Pedantry in quantification and routinization of climatic cure erred in supposing that choosing a residence with an eye to urban topography, exposure to sun, winds and drafts, clothing, and diet would automatically be translated into a cure. In fact, such pedantry was an admission on the part of medical supervisors that the chances of cure were slight and, as Beddoes had intuited, the regimen comprised an amusement with a sheen of theory.[118]

Bloxam's ultimate source of authority was tradition and common sense. In his view, medical travel was justified by its very popularity, unassailable by "expert" attacks from outsiders like Burgess. Madeira's equability was real because it was based on popular belief. It was proven by the patronage of European lung patients. If any of them experienced more pain than on the Continent, this was not because of Madeira's dampness or the *leste*, but their physician's mistake in sending them on an journey into their "last haven of health" when a disease had already reached its terminal stage. Yet in casting blame on science and incompetence, Bloxam insisted that he did not want to be understood as submitting a better scientific account. That would make his position contradictory. Rather, the goal was to show the "insufficiency of the grounds on which other people have arrived at conclusion about Madeira's climate. I have disputed the accuracy of data, and the validity of reasoning. Neither do I wish to assert as a fact really ascertained that Dr Mason's cottage was so damp or so ill chosen a place for his hygrometrical experiments."[119]

Madeira's therapeutic reputation suffered within a decade since this polemic took place. Once a consumption winter resort par excellence, patients and doctors increasingly stayed away from it to the point at which the island ceased to be recommended even for the conditions for which it was known to be safe. Its seasons still counted among the most equable, but its relaxing humidity was inferior to tonic and dry resorts such as Mentone or Egypt, or to the increasingly popular Swiss town of Davos.[120] Doctor Lindsay observed "the modern drift of medical opinion in favour of the view that [the marine resorts of the Mediterranean and the Atlantic] afford on the whole less favourable results in consumption than the mountain sanatoria and the dry inland resorts." Statistical reports indicated its tendency to aggravate consumption. The authoritative Doctor R. E. Scoresby-Jackson referred to the "thousands" that had perished due to the misconceptions about climatotherapy in the south.[121] A report by an authority on consumption showed that out of forty-seven confirmed cases who went to Madeira, thirty-two died within six months of arrival, and the rest shortly afterward.[122] Scientific detractors in popular publications made a plea for dissociating climatotherapy from travel, making it possible to give advice without risking the ascription of ignorance or pandering to fashion. The island's physicians, however, worked to show that indiscriminate advice could not undermine the resort's merits, at least, in relation to milder conditions. But the scientific approach to climatotherapy eventually came to stand for a long-overdue "audit" of views that concealed idiosyncrasy, anecdote, and subjectivity.[123]

Scientific unilateralism continued to endorse a holistic perspective. Some of its champions had no qualms about asserting that there was more to the Mediterranean than Mason's hygrometers would have one believe. Medical topographers also believed that the meteorological features of resorts could not be used to calculate their physiological effects; when "we begin to apply these principles to cases, we are beset with the difficulty of estimating that almost incalculable *quantum*—the individual response to climate."[124] The peripatetic invalid and the lay critic were in this sense not unscientific in supposing that "objective findings" and disciplined regimen could not amount to a cure, nor that they would stop a hyperbolic adulation of the region's virtues. The "mythologists" of the Mediterranean knew that medical journeying served broader aspirations that, in their conjoint influences, might constitute Lindsay's "incalculable *quantum*." Change of air was a change of scene and mental habitus because there was more to a dusty promenade, unsanitary accommodation, miasmatic catacombs, and depressing *sirocco* than met the scientific eye.[125] And why not? Such hazards might have been desirable; they added to the erotic risks of the south as a capricious cultural world nestled amid lush landscapes. The risks might have looked trivial to those who showed "no very obvious signs of illness, [but] who were nevertheless found by their medical advisers, to have some point of natural or acquired weakness."[126]

It should be remembered that expert analyses of resort climates represented more than what the experts stated. The profession's embarrassment with Beddoes's doctor "ignoramus" and the recognition of the complexities involved in climate therapy signaled a consensus about the legitimacy of the practice, but a consensus that required stringent rules of verification. If the previous century's travel to the region opened up a nineteenth-century medical investigation of the region, the Madeira controversy illustrated the ways in which this knowledge revisited the place of its origin, but this time without romance, and without memory of its beginnings. If warm resorts determined the early direction in a theory of change of air, this theory had come of age by the mid-Victorian era to challenge this proposition through climatological and statistical analyses of the Mediterranean. The process was not solely defined by the stress on accountability, precision, and truth. Disputes and discussions about the status of natural knowledge and, by extension, of nature, were more about the actual state of that knowledge or physical conditions: "State of the physical environment serves as a substrate for justifications and rationalization of the way things are, for the demonstration of problems sand their explanation, and for accusations of blame."[127] For example, talk about a proper understanding of climate was also talk about hopes of defeating a disease (Mason), about beliefs in the futility of a change of air from a providential point of view (Burgess), and a vision of therapeutic truth based on the right to illusion as long as it staved off the final hour (Bloxam). More generally, the censure of Mediterranean climate often served as a subterfuge for a Puritanical assault on sexual license and dissipation that the region was perceived to inspire among sun-stricken Britons.[128]

Consciously or not, the criticism was bound to boost home resorts, which during the nineteenth century attracted increasing numbers of middle-class clientele. Depressing statistics about the climatorial "El Dorado," the moral and economical downsides of a southbound journey, and the theodicy of native weather could not but help medical and other businesses in home spas.[129] With the new etiology of consumption as a disease of debility, the cold became "hot." Advertisers for British resorts enlisted hydrometeorology to vouch for the benign thermodynamics of the Gulf Stream as a redeeming factor in an otherwise boreal and wet climate. They regretted the error "engraven in the public mind" that the farther one heads north, the colder winters become and summers increasingly cloudy.[130] "We hold that in our own island may be found climate suitable, ay, much more suitable to those who suffer this dire infliction of Providence [consumption]. In the mild climate of Devonshire, on our own southern coast also, and particularly in the Isle of Wight, let the patient seek for air, it is ever to be found."[131] The work of physical geographers could be quoted because it showed that the plants killed in Middlesex winters could *flourish* throughout the Scottish winter. Such upbeat accounts and meteorological studies of local resorts augmented a patriotic climatotherapy that by the mid-Victorian era successfully vied with the literature on foreign spas.[132] Invalids were encouraged to visit Eastbourne (immune to zymotic disease), Dawlish (good for consumptives arriving there in the winter months), Saltburn-by-the-Sea (whose air combated lethargy), Chelsea and Brompton (good for bronchial invalids), Aberystwyth (reenergizing for those affected by wear and tear).[133] Margate was claimed to have less sickness and fewer deaths than any place in Europe, to which claim a conscientious MD added a list of municipal pitfalls that marred this impression: "open privies, stinking piggeries, neglected stables and stable yards, slaughter houses and their garbage."[134]

The popularity of domestic resorts derived also from economic rationale: The south of England was cheaper and was the home of physicians who had reasons to do up the local weather. By midcentury, domestic tourists found that English seaside, while lacking Italian vistas and glorious sunshine, still offered respite from the pollution, bustle, and fatigue of London and industrial towns. This clientele naturally expanded to the middling classes and wealthier artisans who indulged in the contrived fantasy of "fairy places," as Thomas Hardy described contemporary Bournemouth. Maritime watering places became the manifestation of an imaginary English arcadia,[135] which even the well-off eagerly swapped for the Mediterranean: "An opportunity is now offered of establishing a real Montpellier on the south coast of England and [...] something better than a Montpellier in point of beauty, for the upper and wealthier classes of society [...] ought to be encouraged to remain at home and spend their income in husbanding their health in England."[136] Yet if these and similar invitations flattered patriotic sentiments, the better informed considered them unrealistic. A reviewer of George S. Hooper's medical topography of the Isle of Jersey warned that the "author cannot escape the charge of partiality in favor of his own place of residence." While Jersey possessed a relatively uniform climate year round, its villages, damp and "exposed to every wind," were stricken by catarrhs, bronchitis, and rheumatism that were aggravated by the short-term variability of the atmosphere: "[O]ne day wearing as if by anticipation, the cheering attractions of the summer sky, whilst the next, cold dark and rainy turns the disappointed mind back to the irksomeness of an inclement season. By reasons of these variations, the fashion of low dresses, short sleeves, and bare legs, for children, is admissible on no sound principles, in this island."[137]

By and large, however, trends in consumption treatment used less climatology, despite the drive to warm up and beautify local resorts. Simultaneously, the rise of bacteriology and the open-air treatment at sanatoria announced both a demise of the health-based Grand Tour and the value given to air as such.[138] So did the emphasis on sanitarian negligence as a cause of "climatorial" disease, and many of the English found continental resorts lacking in this respect.[139] But while a growing discrepancy between expectations and real conditions might account for the negative picture of the south, later commentators sometimes emphasized ephemeral reasons. William Gordon, climatologist and physician to the Royal Devon and Exeter Hospital remarked, "Once upon a time the medical faculty of this country laid much stress on climate in the cure of consumption. Now the fashion has changed. For (softly be it spoken) there is a fashion in therapeutics not so very far removed from a fashion in frocks!"[140] What made the medical Mediterranean appealing a century before Gordon was anecdotal evidence, peripatetic culture suffused with notions of elite *Bildung* and the flight from urban civilization.

Conclusion

In this book I have established relationships between medical theories about environmental exposure and British middle-class life. I have argued that such relationships can be identified in several domains in which social practice and expectations shaped medical analyses on the effects of external stressors on bodies and the population at large. During the last third of eighteenth century, engagement with Hippocratic "airs, waters, and places" and Galenic "six things non-natural" acquired a quality of social and moral critique that was latent but not expressly stated in traditional statements of the doctrines. Luxury became the proximate cause of illness, for which therapy involved nonstandard approaches ranging from ventilation to travel to respirators to flannel. Informing these analyses was the guiding idea that external "influence" and bodily "exposure" derived from cultural perceptions of domestic space in more affluent households. When medical and social commentators referred to smell, smoke, clutter, cold, and bad weather, they usually took for granted the aesthetic and hygienic criteria spun around a middle-class sensibility to "discomfort," which "delicate" individuals identified in places untouched by intervention, manipulation, and control. But I would argue that to say such discomforts came about as a result of an elimination of *real* conditions and to make physical space the deliberate target of hygiene would be to commit a "welfare fallacy" based on the retrospective rhetoric of legitimation established by public health reformers in the nineteenth century. I propose instead that the sensibility of eighteenth-century society was the prime source of the production of modern environmental risk.

I have tried to show that that the arguments put forward to legitimize ventilation, dress reform, and travel therapy emerged together with recognition of the possibility of overcoming the pitfalls of existing customs. Cold, damp, cleanliness, and overcrowding became problems only in contrast to their alternatives. Medical reasoning in these discussions was predicated on the notion of exposure and always proceeded on the assumption that solutions were within reach of human ingenuity. What made physical surroundings increasingly problematic was not their objective "nature"—no matter how horrendous they might have been to contemporaries—but their perception as the exact opposites to the "improved" spaces of middle-class households and their immediate vicinity. Intolerability was proportional to a failure to improve since even the possibility of prevention increased the perception of intolerability. As recent scholars have argued, it is useful to think of the notion of causation as linked to our "manipulative techniques for producing results": The early nineteenth-century emergence of "poverty" as a preventable evil came about only in the wake of the growing potential of industry to produce accessible goods. The more one is able to manipulate one's surroundings, the less willing one is to tolerate the persistence of "preventable" evil. But it should be noted that the very perception of preventability was parasitic upon trust in manipulation and intervention. We have seen this in the emergence of the "preventable" evil of bedroom air in the wake of the commercialization of domestic ventilation.[1]

Exposure to a perceived health hazard in this framework was seen as a liability because it revealed a failure of prevention through lack of means or sheer ignorance. Exposure to external threat was an embarrassment in the sense that it signaled a professional and personal incompetence in meeting the expectations implied in a rational system of regimen, hygiene, and other measures used to secure daily health and comfort. Conditions of living space were conducive to health only if they clearly showed the marks of maintenance in line with notions of respectability and comfort. Private space was a matter of medical improvement by means environmental intervention precisely because such improvement could be acquired through menial work, technological gadgetry, and resource-consuming personal regimens. Medical commodification of the domestic realm was in this sense a *status* space marked by a conspicuous waste of labor, wealth, and time.

In the medical literature and practice of the late eighteenth century, maintenance-intensive spaces that were designed with the express intention for healthy dwelling—personal clothing climates, servant-supported households, gardens, select urban areas, smaller neighborhoods and suburbia, even hospitals and Parliament buildings—appeared as both paragons of biological desirability and targets of emulation. In health discourse, however, they were rarely linked to affordability, work, and technical know-how, but to "natural" standards of cleanliness, comfort, and convenience. In health literature, time-, labor-, and finance-intensive infrastructures required to achieve such standards were either left unmentioned or expenditure on them was assumed to be worthy of the effort. Thus "exposure physiology" and the ideology of delicate sensibility ratified these socioeconomic standards as biological facts. In this naturalization, it was a systemic concealment of economic means invested in the production of such spaces that performed a central ideological function in the "invention" of modern healthy space.

Ideas about health *assumed* wealth and labor, but such wealth and such labor were rarely intended for hygienic improvement as such. Rather, the norms of spatio-medical desirability that domestic labor, technology, and leisure time made the sine qua non of middle-class well-being were aimed at keeping up a decorum of class even if the results of this process could result in better hygiene, comfort, and health. As Thornstein Veblen pointed out a century ago, "[T]he effects are pleasing to us chiefly because we have been taught to find them pleasing [and] the housewife's efforts are under the guidance of traditions that have been shaped by the law of conspicuously wasteful expenditure of time and substance. If beauty or comfort is achieved—and it is more or less fortuitous if they are—they must be achieved by means and methods that commend themselves to the great economic law of wasted effort."[2] It was this manufactured propriety and the ritual care of material accoutrements in indoor space by middle-class households that made cleaning, airing, heating, ironing, bleaching, and dusting necessary ingredients of a decent life and—by degrees—the social origins of modern notions of a healthy environment.

I have suggested that early nineteenth-century living space could acquire pathological meaning only through a dialectical construction of physiological vulnerability and external threat in a commercialized context of health literature, technological innovation, and high-end alternative therapies. Here I would like to add that a medicalized space could not even *exist* prior to physiological notions of sensibility, susceptibility, and vulnerability. Neither could the latter exist before there was a consensus about the reality of external threat. In all of these analyses, however, one should note that the idea of externality and external "influence" were value-laden terms that presupposed a particular makeup of individuals on which such an influence could be exerted in the first place. Neither external influences nor surroundings nor "enviroment" can exist as such: all such notions are relational. The larger the quest for personal ease, the riskier the surroundings that were excluded. Exposure and shelter embodied this dialectic in seeing health as a result of planned "positioning" in relation to a constellation of pathogenic milieus.

The space that such a construction of health implied was one in which society was on the defensive. It is possible to take this characterization literally, as in James Riley's phrasing of eighteenth-century medicine as a campaign to "avoid" disease. More pertinently, however, medical space as a *plenum* of influences can be interpreted as a *fetish* generated by anxiety about urban middle-class life associated with the domestic realm and regimen. The maintenance of such status, in other words, *implied* an assumption of external threat. It also *invited* a system by means of which such status could be guarded, augmented, and treated as a natural and inalienable right. It is the contention of this book that such a system was accomplished by a medical theory of exposure, discomfort, and genteel sensibility. Such a "culture of narcissism" is prevalent in societies in which the perceived value of possessions is so high that they require defensive action at all times and so naturalize possession by essentializing the purported threat. In such a society, every action becomes defensive, because the threats to wealth, health, and looks can become so numerous and so opaque that they must remain incomprehensible except when summarily conceptualized as "environment."[3]

In this way a physico-pathological construction of "environment" legitimized middle-class understanding of health as an achievement rather than a baseline. To be able to assert that one's condition could change with a change in "environing" weather or bedroom air, one had to understand the human constitution as something that did not support bad weather, wet clothes, or oppressive atmosphere. In other words, a "zone of health" of one such constitution was a zone free from the effects of such factors. This was a socio-medical space created through control over the contingency of nonowned and therefore nonmaintained space. The construction of atmospheric hazard as a random external threat represented a middle-class medicalization of nonowned, public space. But whether the notion of environmental health depended exclusively on a separation of the private from the public realm remains to be seen. It is nonetheless true that just as nineteenth-century suburbia prospered when a critical number of influential households made it possible for improvement beyond their immediate vicinity, so the nineteenth-century medical profession and the health-conscious classes extrapolated their own sense of discomfort into a full-blown public realm of contingency. "The actual state of the atmosphere, in the habitation of the extremely poor, is too bad and too revolting to be understood by any except those who actually *visit* them."[4] Such statements linked middle-class sensibilities to socioeconomic conditions and provided stock tropes that public health reformers took for granted in their assessment of the "environmental" plight of the poor. Crucially, however, these spatio-medical pairings were eventually dislodged from their original socioeconomic circumstances so that *any* space and *any* location could become hazardous if left unattended.

The medical invention of atmospheric risk was thus one of many fetishes brought about by demographic and economic changes in eighteenth- and nineteenth-century society. "Environmentalism" as a key trope in British medical and social thought during the nineteenth century and beyond was in this sense a reification of elite norms of the quality of living space, a domestication of contingency by means of infrastructural and personal regimen to control exposures.

Notes

Introduction

1. Pilloud and Louis-Courvoisier, "The Intimate Experience of the Body in the Eighteenth Century: Between Interiory and Exteriority," *Medical History* 47 (2003): 451–72, 458. See also Mieneke te Hennepe, *Depicting Skin: Visual Culture in Nineteenth-Century Medicine* (Wageningen: Ponsen and Looijen, 2007).

2. Caroline Hannaway, "From Private Hygiene to Public Health: A Transformation in Western Medicine in the Eighteenth and Nineteenth Centuries," in Ogawa (ed.), *Public Health: Proceedings of the 5th International Symposium on the Comparative History of Medicine* (Tokyo: Saikon Publishing Company, 1981), 108–28.

3. Anne Buttimer, "Airs, Waters, Places: Perennial Puzzles of Health and Environment," in Nicolaas Rupke (ed.), *Medical Geography in Historical Perspective*. Medical History, Supplement No. 20 (The Wellcome Trust Centre for the History of Medicine at UCL: London, 2000). On Hippocratic environmentalism see Frederick Sargent II, *Hippocratic Heritage: A History of Ideas about Weather and Human Health* (New York: Pergamon Press, 1982).

4. Alain Corbin, *The Foul and the Fragrant: Odor and the French Social Imagination* (Lemington Spa: Berg, 1986).

5. Caroline Hannaway, "The Societe Royale de Medecine and Epidemics in the Ancient Regime," *Bulletin of Medical History* 46 (1972): 257–73; Ludmilla Jordanova, "Policing Public Health in France, 1780–1815," in Teizo Ogawa (ed.), *Public Health* (Tokyo: Taniguchi Foundation, 1981), 12–32; Henry Lowood, *Patriotism, Profit and the Promotion of Science in the German Enlightenment* (New York and London: Garland 1991); Mary Lindemann, *Health and Healing in Eighteenth-Century Germany* (Baltimore and London: John Hopkins University Press, 1996), 271–72.

6. Francis Lobo, "John Haygarth, Smallpox, and Religious Dissent in Eighteenth-Century England," in Andrew Cunningham and Roger French (eds.), *The Medical Enlightenment of the Eighteenth Century* (Cambridge University Press, Cambridge, 1990), 217–53; Christopher

Lawrence, "Disciplining Disease: Scurvy, the Navy, and Imperial Expansion, 1750–1820," in David Phillip Miller and Peter Hans Reill (eds.), *Visions of Empire: Voyages, Botany, and Representations of Nature* (Cambridge: Cambridge University Press, 1996), 80–106.

7. Hans-Joachim Voth, "Time Use in Eighteenth-Century London: Some Evidence from the Old Bailey," *The Journal of Economic History* 57, No. 2 (1997): 477–99.

8. Simon Schaffer, "Measuring Virtue: Eudiometry, Enlightenment and Pneumatic Medicine," in Andrew Cunningham and Roger French (eds.), *The Medical Enlightenment of the Eighteenth Century* (Cambridge: Cambridge University Press, 1990), 281–318; on travel and health, David Arnold (ed.), *Warm Climates and Western Medicine: The Emergence of Tropical Medicine, 1500–1900* (Rodopi: Amsterdam, 1996); on early studies in altitude physiology, see Fergus Fleming, *Killing Dragons: The Conquest of the Alps* (London: Granta, 2000); on travel, see Christopher Hoolihan, "Health and Travel in Nineteenth-Century Rome," *Journal of the History of Medicine and Allied Sciences* (1989): 462–85; on various environmental measures, see James Riley, *The Eighteenth-Century Campaign to Avoid Disease* (New York, 1978). On the early history of the umbrella, see T.S. Crawford, *A History of the Umbrella* (David and Charles: Newton Abbot, 1970); suburban England is the subject of Carl Estabrook, *Urbane and Rural England: Cultural Ties and Social Spheres in the Provinces* (Manchester: Manchester University Press, 1998). On sports, see Henning Eichberg, "The Enclosure of the Body—On the Historical Relativity of 'Health,' Nature and the Environment of Sport," *Journal of Contemporary History* 21 (1986): 99–121; on cemeteries, see Thomas W. Laqueur, "The Places of the Dead in Modernity," in Colin Jones and Dror Wahrman (eds.), *The Age of Cultural Revolutions: Britain and France, 1750–1820* (University of California Press: Berkeley, 2002), 17–32.

9. L.J. Jordanova, "Earth Science and Environmental Medicine: The Synthesis of the Late Enlightenment," in L.J. Jordanova and Roy S. Porter (eds.), *Images of the Earth: Essay in the History of the Environmental Sciences* (BHSH: London, 1978), 119–46. The terminology is discussed in James Riley, *The Eighteenth-Century Campaign to Avoid Disease* (New York: St. Martin Press, 1987), xv–xvi. David Arnold, *Colonizing the Body: State Medicine and Epidemic Disease in Nineteenth-Century India* (Berkeley: University of California Press, 1993); Mark Harrison, *Climates and Constitutions: Health, Race, Environment, and British Imperialism in India* (Oxford: Oxford University Press, 1999). See also Andrea Rusnock, "Hipocrates, Bacon, and Medical Meteorology at the Royal Society, 1700–1750," in David Cantor (ed), *Reinventing Hippocrates* (Ashgate: Aldershot, 2002), 136–153. Charles E. Rosenberg, "Medical Texts and Social Context: Explaining William Buchan's *Domestic Medicine*," in idem.,

Explaining Epidemics and Other Studies in the History of Medicine (Cambridge: Cambridge University Press, 1992), 32–56.

10. Spiro Kostof, *The City Assembled: Elements of Urban Form through History* (Little Brown, Boston 1992), 132; Mark Jenner, "The Politics of London Air: John Evelyn's 'Fumifugium' and the Restoration." *The Historical Journal*, 38 (1995): 535–51.

11. Michael Dillon, *Politics of Security* (London: Routledge, 1996), 19.

12. Tom Fort, *Under the Weather* (London: Arrow Books, 2006), 12.

13. Jan Golinski, *British Weather and the Climate of Enlightenment* (Chicago: Chicago University Press, 2007), 11.

14. Ian Wilkinson, *Anxiety in a Risk Society* (London and New York: Routledge, 2001).

15. Odo Marquard, *Farewell to the Matters of Principle* (Translated by Robert M. Wallace. Oxford: Oxford University Press, 1989), 11.

16. *Lloyd's Evening Post and British Chronicle*, November 30, 1761.

17. But see Severine Pilloud and Micheline Lois-Courvoisier, "The Intimate Experience of the Body in the Eighteenth Century: Between Interiority and Exteriority," *Medical History* 2003, 47: 451–72.

18. "Environmental Medicine Focuses on the Causes of Disease in an Environmental Context. The Environment Creates Exposures to Many Different Physical, Biological and Chemical Agents," http://www.medterms.com/.

19. Noel Castree, "Environmental Issues: From Policy to Political Economy," *Progress in Human Geography* 26 (2002): 357–65, 357; David Harvey, *Justice, Nature, and the Geography of Difference* (Oxford: Blackwell, 1995), 117–19. On the uses of the term, see K. Eden, "Environmental Issues: Knowledge, Uncertainty, and the Environment," *Progress in Human Geography* 22 (1998): 425–32.

20. Simon Dalby, "Ecological Metaphors of Security: World Politics in the Biosphere," *Alternatives* 23 (1998): 291–319, 295.

21. Harvey, *Justice, Nature, and the Geography of Difference*, 117.

22. Georges Canguilhem, "The Living and Its Milieu", *Grey Room* 3 (2001): 7–31, 8.

23. Sara Warneke, "A Taste for Newfangledness: The Destructive Potential of Novelty in Early Modern England." *Sixteenth Century Journal* 26 (1995): 881–95.

24. Robert John Thornton, *Medical Extracts* (London: J. Johnson, 1798) vol. 2, 267.

25. Hugh Smythson, *The Compleat Family Physician* (London: Harrison and Co, 1785), 108.

26. *True Briton*, September 3, 1799; Thomas Beddoes, *Essay on the Causes, Early Signs, and Prevention of Pulmonary Consumption for the Use of Parents and Preceptors* (Bristol, 1799), 129. William Davidson, *Observations, Anatomical, Physiological, and Pathological, on the Pulmonary System* (London, 1795), 97. *Times,* April 12, 1797; John Mudge, *A Radical and Expeditious Cure for a Recent Catarrhous*

Cough (London: E. Allen, 1780), x; Martin Clare, *The Motion of Fluids, Natural and Artificial* (London: Edward Symon, 1735), 188; *Morning Chronicle and London Advertiser* (March 25, 1788).

27. Charles Bisset, *An Essay on the Medical Constitution of Great Britain* (London: A. Millar, 1762), 1.

1   Exposed and Vulnerable

1. *Daily Advertiser*, March 7 "London Debates: 1789," in *London Debating Societies, 1776–1799* (1994), 246–73.

2. On health and weather diaries, see Jan Golinski, "Barometers of Change: Meteorological Instruments as Machines of Enlightenment," in William Clark, Jan Golinski, and Simon Schaffer (eds.), *The Sciences in Enlightened Europe* (Chicago and London: Chicago University Press, 1999), 69–93. See also F.N.L. Poynter, "Sydenham's Influence Abroad," *Medical History* 17 (1973): 223–234. Main protagonists in this tradition included Clifton Wintringham, *Commentarium nosologicum morbos epidemicos et aeris variationes in urbe Eboracenci locisque vicinis, ab anno 1715, usque ad finem anni 1725* (Londini: J. Clarke, 1727); John Huxham, *Observations of the Air and Epidemical Diseases, Made at Plymouth from 1728–1737* (Translated from Latin. London: J. Hinton, 1759); Charles Bisset, *An Essay on Medical Constitution of Great Britain* (London, 1762); John Rutty, *A Chronological History of the Weather and Seasons and of the Prevailing Diseases in Dublin* (London: Robinson and Roberts, 1770); Thomas Short, *A General Chronological History of the Air, Weather, Seasons, Meteors etc in Sundry Places and Different Times* (London: T. Longman and A. Millar, 1749).

3. John Murray, MD, "Journal Containing Daily Meteorological and Monthly Medical Observations," Wellcome MS.7840. See also "Meteorological Observations [in Bristol, January 1774]," Wellcome Library MS. MSL. 111.

4. Carlo M. Cipolla, *Miasmas and Disease: Public Health and the Environment in the Pre-Industrial Age* (Hew Haven and London: Yale University Press, 1992); on poisonous airs, see John Prestwich, *Dissertation on Mineral, Animal, and Vegetable Poisons* (London: F. Newberry, 1775), 71–106.

5. All things considered to act "immediately upon the Body, or make Impression on it, at a Distance, through the Medium of the Senses, the changes that happen in the Atmosphere which surrounds us, and the Air we breath, constitute the general, external causes that affect Health," Andrew Hamper, *The Economy of Health*, (London: Author, 1785), 1. James Dunbar, *Essay on the History of Mankind in Rude and Cultivated Ages* (London: W. Strahan, 1781), 221–2. For an overview, see Clarence J. Glacken, *Traces on the Rhodian Shore* (Stanford: University of Claifornia Press, 1990), chapter 12. See also

Christopher J. Berry, "'Climate' in the Eighteenth Century: James Dunbar and the Scottish Case," *Texas Studies in Literature and Language* 16 (1974): 281–92.

6. James Mackenzie, *The History of Health* (Edinburgh: W. Gordon, 1758), 368; Andrew Wear, "Making Sense of Health and the Environment in Early Modern England," in Andrew Wear (ed), *Medicine in Society: Historical Essays* (Cambridge: Cambridge University Press, 1992), 119–147. On locality and weather, see Vladimir Janković, "The Nature of Place and the Place of Nature," *History of Science* 38 (2000): 79–113.

7. Gail Kern Paster, *The Body Embarrassed: Drama and the Disciplines of Shame in Early Modern England* (Ithaca, NY: Cornell University Press, 1993), 9, and Barbara Duden, *The Woman Beneath the Skin: A Doctor's Patients in Eighteenth Century Germany* (Cambridge: Harvad University Press, 1998), 119–23, Gianna Pomata, *Contracting a Cure: Patients, Healers and the Law in Early Modern Bologna* (Baltimore: Johns Hopkins University Press, 1998), 132–3.

8. Laurence Brockliss and Colin Jones, *The Medical World of Early Modern France* (Oxford: Clarendon Press, 1997), 462; Alain Corbin, *The Foul and the Fragrant: Odor and the French Social Imagination* (Lemington Spa: Berg, 1986)

9. David Ferriar, "The Erotics of Empiricism," unpublished presentation at University of Leeds, March 2, 2004. "Mental meteorology" is referred to in Barbara Maria Stafford, *Body Criticism: Imaging the Unseen in Enlightenment Art and Medicine* (Cambridge, MA: MIT Press, 1991), 417–36.

10. D. G. C. Allan and R. E. Shoffield, *Stephen Hales: Scientist and Philanthropist* (London: Scholar Press, 1980), 40. This pervasiveness questions Northrop Frye's notion that the eighteenth-century ideas of nature embraced "the physical environment, which human beings are in, but not of," as argued in Northrop Frye, "Varieties of Eighteenth-Century Sensibility," *Eighteenth-Century Studies* 24 (1990): 157–72, 161. See also Percy M. Dawson, "Stephen Hales, The Physiologist," *The Johns Hopkins Hospital Bulletin* 15 (1904): 1–15.

11. John Arbuthnot, *An Essay Concerning the Effects of Air on Human Bodies* (London: J. Tonson, 1733), 2. On Arbuthnot, see David E. Shuttleton, "A Modest Examination: John Arbuthnot and the Scottish Newtonians," *British Journal for the Eighteenth Century Studies* 18 (1995): 47–62; Alexander Sutherland, *Attempts to Revive Antient Medical Doctrines. I. Of Waters in General. II. Of Bath and Bristol Waters in Particular. III. Of Sea Voyages* (London: A. Millar 1763), vol. 2, 132. See also Ebenezer Sibly, *The Medical Mirror* (London, [1800?]). On the issue of necessity of air, see William Forster, *A Treatise on the Causes of Most Diseases Incident to Human Bodies* (London: J. Clarke, 1746), 239.

12. Bernard Lynch, *A Guide to Health through the Various Stages of Life* (London: Cooper, 1744), 134.

13. Lynch, *A Guide to* Health, 92–109; for a taxonomy of ingredients, see John Sinclair, *The Code of Health and Longevity* (3 vols. Edinburgh: A. Constable, 1807), vol. 1, 178–9.

14. Jeremiah Wanewright, *Mechanical Account of the Non-Naturals* (London: Ralph Smith, 1707), 51. Another mechanical approach is John Burton, *A Treatise on the Non-Naturals* (York: A. Staples, 1738). The principle of circulation, perspiration, and digestion, the air was "introduced into all the bodies, that have any kind of life." James Drake's *Anthropologia Nova* (London: Samuel Smith and Benjamin Walford, 1707), 181. On iatromechanism, see Theodore Browne, *The Mechanical Philosophy and the "Animal Economy"* (New York: Arno Press, 1981); Anita Guerrini, "Archibald Pitcairne and Newtonian Medicine," *Medical History* 31 (1987): 70–83.

15. Wainewright, *Mechanical Account*, 62–63.

16. George Cheyne, *The English Malady* (London: G. Strahan, 1733).

17. George Cheyne, *An Essay on Regimen* (London: C. Rivington and J. Leake, 1740), ii.

18. George Cheyne, *A New Theory of Continual Fevers* (H. Newman, and J. Nutt, 1701), 26, 74.

19. George Cheyne, *An Essay on Health and Long Life* (London: G. Strahan and J. Leake, 1724), 47.

20. Cheyne, *An Essay on Health and Long Life*, 115.

21. "Such indeed is the infirmity of human body, that being inefected sometimes with even the slightest stain, either through error or carelessness, it is obnoxious to the greatest danger. Wherefore we may conclude, that he has made no small progress in the knowledge of disease, who has not overlooked the meanest things," John Freind, *Nine Commentaries upon Fevers* (London: T. Cox, 1730), 125.

22. Arbuthnot, *Essay*, 212–24.

23. Arbuthnot conceded that it was "the miracle of providence" that the air remained for the most part healthy to all. This worry spurred several ad hoc theodicies that asked medical professionals to explain the purported dangers. John Huxham, for example, calculated that the body could endure sixteen tons of air, because God had compensated it by equalizing its pressure from all sides. A similar earlier writing on the same subject is Richard Boulton, *Some Thoughts Concerning the Unusual Qualities of the Air* (London: John Hooke, 1724).

24. Charles Rosenberg (ed.), *Right Living: An Anglo American Tradition of Self-Help Medicine and Hygiene* (Baltimore: John Hopkins University Press, 2003) and Ginnie Smith, "Prescribing the Rules of Health: Self-Help and Advice in the Late Eighteenth Century," in Roy Porter (ed), *Patients and Practitioners* (Cambridge: Cambridge University Press, 1985): 249–82.

25. Cheyne, *Essay*, 173.

26. William Coleman, "Health and Hygiene in the Encyclopaedia: A Medical Doctrine for the Bourgeoisie," *Journal of the History of Medicine* (1974): 399–421; see also Antoinette S. Emsch-Deriaz, *Tissot: Physician of the Enlightenment* (New York: Peter Lang, 1992).

27. Baer H., Singer, M. and Johnsen, J., "Toward a Critical Medical Anthropology," *Social Science and Medicine* 34 (8): 95–98.

28. Lynch, *A Guide to Health*, 96, 136.

29. Brown, *The Mechanical Philosophy*, 345–47.

30. Roselyne Rey, "Vitalism, Disease and Society," in Roy Porter (ed.), *Medicine in the Enlightenment* (Amsterdam: Rodopi, 1995), 278; Lester King, "Some Problems of Causality in 18th Century Medicine," *Bulletin of the History of Medicine* 37 (1963): 15–24.

31. Duden, *The Woman Beneath the Skin*, 14. Christopher Lawrence, "The Nervous System and Society in the Scottish Enlightenment," in Steven Shapin and Barry Barnes (eds.) *Natural Order* (London: Sage Publications, 1979), 19–40. See also George Rousseau, "Nerves Spirits and Fibres: Towards Defining the Origins of Sensibility." *The Blue Guitar* 2 (1976): 125–53; Sergio Moravia, "From Homme Machine to Homme Sensible: Changing Eighteenth-Century Models of Man's Image," *Journal of the History of Ideas* 39 (1978): 45–60.

32. Louis La Caze, *Idee de l'homme physique et moral* (Paris: H.L. Guerin and L.F. Delatour, 1755). See Elizabeth A. Williams, *The Physical and the Moral: Anthropology, Physiology, and Philosophical Medicine in France, 1750–1850* (Cambridge: Cambridge University Press, 1994), 41–45.

33. Williams, *The Physical and the Moral*, 153. Hygienists who rendered the "environment [...] social in nature," see William Coleman, *Death Is a Social Disease: Public Health and Political Economy in Early Industrial France* (Madison: Wisconsin University Press, 1982), 202.

34. Brockliss and Jones, *The Medical World of Early Modern France*; Pedro Entralgo, *Mind and Body: Psychosomatic Pathology: A Short History of the Evolution of Medical Thought* (London: Harvill, 1955); Literature on sensibility and the nerves is now extensive and varied, some of which is annotated in Felicity Nussbaum and Laura Brown (eds.) *The New Eighteenth Century* (London: Methuen, 1987). See also Anne Vila, *Enlightenment and Pathology: Sensibility in the Literature and Medicine of Eighteenth-Century France* (London: The Johns Hopkins University Press, 1998); Markman Ellis, *The Politics of Sensibility: Race, Gender and Commerce in the Sentimental Novel* (Cambridge: Cambridge University Press, 1996).

35. George Rosen, "The Fate of the Concept of Medical Police," *Centaurus* 5 (1957): 99; On the nonnaturals as necessary for life, see Luis Garcia-Ballester, *Galen and Galenism* (London: Ashgate/ Variorum, 2002), "On the Organs of the 'Six Non-Natural Things'"

in Galen (117–152). "[I]n reference to health and sickness, the influence of the air on the human bodies is very surprising; for although a man may and can exist for three days, without taking the usual refreshments to nourish and to support nature, yet the same man could barely live a moment without air: in short the existence of men depends on their enjoyment of air, which is absolutely necessary for life, etc." Jerome J. Bylebyl, in "Galen on the Non-Natural Causes of Variations in the Pulse," *Bulletin of the History of Medicine* 45 (1971): 482–85, writes: "Not by nature, in medicine, designated those things (*res non naturales* in medieval Latin terminology) which, though necessary for life, were not part of man's natural endowment." See also L.J. Rather, "The Six Things Non-Natural: A Note on the Origins and Fate of a Doctrine and a Phrase," *Clio Medica* 3 (1968): 337–347. On French popularization, see Antoniette Emsch-Deriaz, "The Non-Naturals Made Easy," in Roy Porter (ed.), *The Popularization of Medicine* (London and New York: Routledge, 1992), 134–59. See also *Hippocrates's Treatise on the Preservation of Health.* (London: John Bell, 1776) and Francis Clifton, *Hippocrates UPON Air, Water, and Situation* (London: J. Watts, 1734); Saul Jarcho, "Galen's Six Non-Naturals: A Bibliographic Note and Translation," *Bulletin for the History of Medicine* 44 (1970): 372–77; Peter H. Niebyl, "The Non-Naturals," *Bulletin for the History of Medicine* 45 (1971): 486–92. Charles T. Wolfe and Motoichi Terada, "The Animal Economy as Object and Program in Montpellier Vitalism," *Science in Context* 21 (2008): 537–79.

36. Robert Gusthart, *Specimen Medicum Inugurale de Aere Ejusque in Respiratione usu et Effectibus* (Edinburgh: Thomas Ruddimann, 1740); Robert Willan *Dissertatio Medica Inauguralis de Qualitatibus Aeris* (Edinburgh: James Cheyne, 1745); Ebenezer Macfait, *Dissertatio Medica Inauguralis de Aere, Aquis et Locis* (Edinburgh: 1745); John Gowdie, *Dissertatio Medica Inauguralis, de Aere Quatenus Morborum Causa* (Edinburgh: Murray and Cohran, 1754); James Johnstone, *Tentamen Medicum Inaugurale de Aeris Factitii Imperio in Primis Corporis Humani Viis* (Edinburgh: T. and W. Ruddimannos, 1750); William Brown, *Specimen Inaugurale Pathologicum, de Viribus Atmosphaerae Sentienti Obviis* (Edinburgi: Apud Balfour, Auld, et Smellie, 1770), Daniel Rutherford, *Dissertation Inauguralis de Aere Fixo Dicto, Aut Mephitico, etc.* (Edinburgh: Balfour and Smellie, 1772); Edmundus Cullen, *Tentamen Medicum Inugurale de Aere, et Imperio Eju in Corpora Humana.* (Edinburgh: Balfour and Smellie, 1781); Henri Burton, *Tentamen Physiologico-Medicum de Usu et Effectu Aeris Puri in Corpus Humanum.* (Edinburgh: Balfour and Smellie), 1788; Henry Robertson, *Dissertatio Chemica Medica Inauguralis, de Aere Atmosphaerico* (Edinburgh: C. Stewart, 1801); William Cheekes, *Disputatio Chemica Inauguralis de Aere* (under George Baird) (Edinburgh: Adam Neill and Co., 1803); Nicholas

Pitta, *Dissertatio Physiologica Inauguralis, de Caeli Effectu in Genus Humanum* (Edinburgh: C. Stewart, 1812); George Samuel Jenks, *Dissertatio Medica Inauguralis de Coelo Tabescentibus Benigno* (Edinburgh: J. Pillans, 1821); William Jackson, *Tentamen Chemica Medica Inauguralis de Aëre Communi* (Edinburgh: P. Neill, 1822). For a general treatment of the subject, see Lisa Rosner, *Medical Education in the Age of Improvement: Edinburgh Students and Apprentices, 1760–1826* (Edinburgh: Edinburgh University Press, 1991).

37. Thomas Broman, "The Medical Science," in Roy Porter, *The Cambridge History of Science: Eighteenth-Century Science* (Cambridge: Cambridge University Press, 2003), 463–84. Treatises on regimen, health preservation, and nonnaturals rose in popularity: William Buchan's *Domestic Medicine: Or, a Treatise on the Prevention and Cure of Diseases by Regimen and Simple Medicines* (London: W. Strahan, 1772) went through twenty editions by 1797.

38. Christopher Lawrence, "The Nervous System and Society in the Scottish Enlightenment," 24. Another domain in which the Scots exerted influence is discussed in B. White, "Scottish Medicine and the English Public Health," in Derek Dow (ed.), *The Influence of Scottish Medicine* (Carnfort: Pantheon, 1988), pp. See also David Hamilton, *The Healers: A History of Medicine in Scotland* (Edinburgh: Canongate, 1981); Vern and Bonnie Bullough, "The Causes of the Scottish Medical Renaissance of the Eighteenth Century," *Bulletin for the History of Medicine* 45 (1971): 13–28.

39. Sharon Ruston, *Shelley and Vitality* (London: Palgrave, 2005), Andrew Burstein, "The Political Character of Sympathy," *Journal of the Early Republic* 21 (2001): 601–632 .

40. Rosalyn Rey, "Vitalism, Disease and Society," *Clio Medica* 29 (1995): 274–288.

41. Christopher John Lawrence, "Medicine as Culture: Edinburgh and the Scottish Enlightenment," (PhD thesis, University College London, 1984), 143.

42. Lawrence, "Medicine as Culture," 145 and chapters 7 and 8.

43. Lawrence, "The Nervous System," 29. Note Cullen's personification of climate as "rude and uncultivated," linking the shaper with the shaped."

44. G.J. Barker-Benfield, *The Culture of Sensibility* (Chicago: University of Chicago Press, 1996), 8–9.

45. Barker-Benfield, *The Culture of Sensibility*, 9. On the memory of place, see Severine Pilloud and Micheline Louis-Courvoisier, "The Intimate Experience of the Body in the Eighteenth Century: Between Interiority and Exteriority," *Medical History* 47 (2003): 451–472; hereditary climate applied to the more common notion of "native clime" and is used by Thomas Burgess, "Inutility of Resorting to the Italian Climate for the Cure of Pulmonary Consumption," *Lancet*

55 (1850): 591. Daniel Cottom, "In the Bowels of the Novel: The Exchange of Fluids in the Beau Monde," *Novel: A Forum on Fiction* 32 (1999): 157–86.

46. George Rousseau "Nerves Spiritis and Fibres: Toward Defining the Origins of Sensibliity," in R.F. Brissenden and J.C. Eade (eds), *Studies in the Eighteenth Century III: Papers Presented at the Third David Nichol Smith Memorial Seminar, Canberra, 1973* (Toronto: University of Toronto Press, 1976).

47. John Gregory, *A Father's Legacy to His Daughters* (Edinburgh: A. Strahan and T. Cadell, 1774), 10.

48. James Gregory, "Theory of Medicine, Notes of Lectures" (Edinburgh, ca. 1785), Wellcome Trust Library MS 2597.

49. Katharine Anderson, "Instincts and Instruments," in Christopher D. Green, Marlene Shore and Thomas Teo (eds.), *The Transformation of Psychology: Influences of 19th-Century Philosophy, Technology, and Natural Science* (Washington: American Psychological Association, 2001), 165. Where the robust handled almost any change, "the infirm cannot well bear the change, and therefore ought to be more careful as to these matters," Huxham, *Observations*, iv.

50. John Forthergill, 1753, quoted in Thomas Willan, *Reports on the Diseases in London Particularly during the Years 1796, 1797, 1798, 1797, and 1800* (London: Phillips, 1801), 242–43. The old and delicate, explained Beddoes in 1806, risked dangers if they went out on cold days. These were the times when senior valetudinarians should stay "wrapped in a six month torpor with the marmot and the doormouse. Let them not be frozen during the sleep. Let them keep their bedrooms at a temperature between 50 and 60 and not allow themselves to imagine that *they* are to be braced by the rigors of a wintry night," Thomas Beddoes, *Manual of Health: Or the Invalid Conducted Safely through the Seasons* (London: J. Johnson, 1806), 46–7.

51. Samuel Johnson wrote: "[A]s politeness increases, some terms will be considered as too gross and vulgar for the delicate." Samuel Johnson, *A Dictionary of the English Language* (London: W. Strahan, 1755–56), preface.

52. David Fordyce, *Dialogues Concerning Education* (London: E. Dilly, 1757), vol. 2, 197.

53. Robert Whytt, *An Essay on the Vital and Other Involuntary Motions of Animals* (Edinburgh: Hamilton, Balfour and Neill, 1751), 46.

54. Robert Whytt, *Observations on the Nature, Causes, and Cure of Those Disorders Which Have Been Commonly Called Nervous, Hypochondriac, or Hysteri* (Edinburgh: T. Beckett, 1756), 307.

55. Whytt, *Observations*, 347–48.

56. Mr Phelps, *The Human Barometer: Or the Living Weather Glass. A Philosophic Poem* (London: M. Cooper, 1743), Preface. "Connections strange, Body and Soul conjoined! That Senseless

Matter should inform the Mind!" Ibid., 15. Terry Castle, *The Female Thermometer: Eighteenth-Century Culture and the Invention of the Uncanny* (Oxford: Oxford University Press, 1995), Jan Golinski, "The Human Barometer: Weather Instruments and the Body in Eighteenth-Century England," Paper given at the American Society for Eighteenth-Century Studies Annual Meeting, Notre Dame, Indiana, April 3, 1998. Cheyne explained that the English climate, which in Voltaire's view was so dull as to breed misanthropy, sedentariness, growth of towns, and luxury "have brought forth a class of distemper with atrocious and frightful symptoms, scarce known to our ancestors," cited in Oswald Doughty, "The English Malady of the Eighteenth Century," *The Review of English Studies* 2 (1926): 257–69, 259.

57. John Millar, *Observations on the Prevailing Diseases in Great Britain* (London: T. Cadell), 290.

58. Robert Thornton, *The Philosophy of Medicine* (London: C. Whittingham, 1799), 342.

59. Benjamin Hutchinson, *Biographia Medica* (London: J. Johnson, 1799), 317.

60. Charles Perry, *Mechanical Account and Explication of the Hysteric Passion* (London: Shuckburgh, 1755), 5.

61. Richard Brocklesby, *Reflections on Antient and Modern Musick, With the Application to the Cure of Diseases* (London: M. Cooper, 1749), 52.

62. Quoted in George Sigmond, "Materia Medica and Therapeutics," *Lancet* 30 (1837) (390–95): 393

63. *True Briton* (March 27, 1793); *Oracle and Public Advertiser* (March 13, 1795); *Telegraph* (July 11, 1796); *Sun* (September 20, 1798); *Sun* (December 28, 1799); *E. Johnson's British Gazette and Sunday Monitor* (December 11, 1796); *Johnson's British Gazette and Sunday Monitor* (December 11, 1796); *Morning Post and Daily Advertiser* (October 25, 1776); *Stuart's Star and Evening Advertiser* (March 24, 1789).

64. Whytt, *Observations*, 25; For a more detailed discussion on Whytt's work on sympathy and its relevance for the development of "reflex," see Ruth Leys, *From Sympathy to Reflex: Marshall Hall and His Opponents* (Garland: New York, 1990), 40–41. On Whytt, see William Seller, "Memoirs of the Life and Writings of Robert Whytt," *Transactions of the Royal Society of Edinburgh* 23 (1864): 99–131; R.K. French, *Robert Whytt, The Soul and Medicine* (London, 1969).

65. James Makittrick Adair, *Commentaries on the Principles and Practice of Physic* (London: T. Becket, 1772), 26.

66. Adair, *Commentaries*, 25.

67. Adair, *Commentaries*, 39.

68. Adair, *Commentaries*, 93; see also Seguin Henry Jackson, *A treatise on Sympathy* (London: J. Murray, 1781); earlier discussions include

Thomas Willis, *The Anatomy of the Brain* (London: T. Dring, 1681), 157; on "secret consent and sympathy," and the interstinal-cutaneous sympathy mentioned by Hippocrates, see Girogio Baglivi, *The Practice of Physick, Reduc'd to the Ancient way of Observations Containing a Just Parallel between the Wisdom and Experience of the Ancients,* (London: Andrew Bell, 1704), 13, 277–79; on intestinal sympathy and the sensitivity of the upper stomach, see James Keill, *The Anatomy of the Humane Body Abridg'd* (London: Ralph Smith, 1703), 36; for astro-medical readings, see Nicholas Culpeper, *The English Physician Enlarged* (London: A. and J. Churchill, 1708), 362.

69. John Leake, *A Course of Lectures on the Theory and Practice of Midwifery* (London: A.D. 1767); idem, *Medical Instructions Towards the Prevention and Cure of Chronic Diseases Peculiar to Women* (London: R. Baldwin, 1777).

70. Leake, *Medical Instructions*, 365.

71. Leake, *Medical Instructions*, 374.

72. Leake *Medical Instructions*, 385, 386, 388.

73. Leake, *Medical Instructions*, 380.

74. Leake, *Medical Instructions*, 385. During the intense cold, on the other hand, the freezing air repelled the blood to the heart and brain and caused sleepiness, stupor, and death. Leake observed that the reason people sleep better in the country owed to the greater gravity of the "air acting externally on the body," Ibid, 391.

75. Leake, *Medical Instructions*, 396.

76. Simon Schaffer, "Enlightened Automata," in Clark, Golinski, and Schaffer (eds.). *The Sciences in Enlightened Europe,* 158. See also Mary Terrall, "Metaphysics, Mathematics and the Gendering of Science in Eighteenth century France," in ibid, 246–71.

77. *An Essay on the Most Rational Means of Preserving Health* (London: James Wallis, 1799).

78. For case studies on weather conditions in the works on surgery, midwifery, general practice, and therapeutics, see Herman Boerhaave, *Dr. Boerhaave's Academical Lectures on the Theory of Physic,* (6 vols., London: W. Innys, 1742–46), vol. 6, 241; Rice Charleton, *Cases of Patients Admitted into the Hospital at Bath, under the Care of the Late Dr. Oliver* (London: T. Cruttwell, 1776), 63; Thomas Denman, *An Introduction to the Practice of Midwifery* (2 vols., London: J. Johnson, 1794–5), vol. 1, 20; Bartholomew di Dominiceti, *Medical Anecdotes of the Last Thirty Years* (London: L. Davis, 1781), 324; John Hill, *Hypochondriasis. A Practical Treatise on the Nature and Cure of That Disorder; Commonly Called the Hyp and Hypo* (London: T. Trueman, 1775), 36; James Latta, *A Practical System of Surgery* (Edinburgh: G. Mudie, 1793), vol. 3, 324; Anton Freiherr von Störck, *A Second Essay on the Medicinal Virtues of Hemlock* (London: T. Becket, 1761), 88.

79. A.F.M. Willich, *Lectures on Diet and Regimen: Being a Systematic Inquiry into the Most Rational Means of Preserving Health and*

Prolonging Life (London: A. Strahan, 1800), 58–9. On the physiology of capillaries, see William Cadogan, *A Dissertation on the Gout, and All Chronic Diseases* (London: J. Dodsley, 1771).

80. J. Lyons, *Fancy-logy: A discourse on the doctrine of the necessity of human actions, proving it to be a fanaticism* (London: W. Bickertonn 1744), 15.

81. J. Lyons, *Fancy-logy:* 13. [Samuel Strutt], *An Essay Towards Demonstrating the Immateriality, and Free-Agency of the Soul* (London: Shuckburgh, 1740), 135. See also Anthony Collins, *A Philosophical Inquiry Concerning Human Liberty* (London: R. Robinson, 1717); John Locke, *An Essay Concerning Human Understanding* (Edited by P. D. Niditch, Oxford: Clarendon Press, 1975), 233–87; *A Discourse on the Doctrine of the Necessity of Human Actions* (London: W. Bickerton, 1744).

82. Thomas Gordon, *Humorist*, 91.

83. See George Pitcher, "Necessitarianism," *The Philosophical Quarterly* 11 (July 1961): 201–212; Gideon Yaffe, *Manifest Activity: Thomas Reid's Theory of Action* (Oxford: Oxford University Press, 2004), 2–3.

84. *The London Magazine. Or, Gentleman's Monthly Intelligencer* (1747), 21, 393. As Dorothy Porter put it, "the eighteenth century invented a culture of sensibility which invited one to be sick to get ahead." Dorothy Porter, "Healthy Body," in Pickstone and Cooter, *Medicine in the 20th Century* (Taylor and Francis, 2000).

85. Willich, *Lectures on Diet and Regimen*, 26, author's emphasis.

86. Philip Dormer Stanhope Chesterfield, *Letters Written by the Late Right Honourable Philip Dormer Stanhope, Earl of Chesterfield, to His Son, Philip Stanhope* (4 vols., fifth edition. London: J. Dodsley, 1774), Vol. 4, 150. For example, "I am well-nigh choked with the sulfurous heat of the weather—or I am unwell," (1826, Journal of Sir Walter Scott). "Crabbe told us that Lord Chesterfield was the first person who introduced the word 'unwell' into common use, and it was forthwith admitted into the vocabulary of fashion," *Oxford English Dictionary*.

87. *Angeline: Or Sketches from Nature* (3 vols. London: Kerby, 1794), vol. 1, 61.

88. *Arley: Or the Faithless Wife* (2 vols. London: J.S. Barr, 1790), vol. 1, 11. Thomas de Quincey, following Ehregott Andreas Christoph Wasianski biography of Kant, captured the unease in reflecting on the philosopher's "positive" health sensation as going beyond the "absence of pain, and of irritation and also of malaise (either of which, though not pain, is often worse to bear)." Paul Youngquist, "De Quince's Crazy Body," *PMLA* 114 (1999): 346–358; see also Charles J. Rzepka, "De Quincey and Kant," *PMLA* 115 (2000): 93–4.

89. *Augusta Denbeigh* (Dublin: Brett Smith, 1795), 88, 210. Unwell meant "indisposed" or "unable." "[W]hat is the matter (looking

tenderly on her). You are unwell?" asks a character his female friend in another novel, to which she replies, "I feel a little indisposed," *Mariamne, or Irish Anecdotes* (2 vols. Dublin: B. Smith, 1794). Vol. 1, 84. Or: "My mother is really unwell [...] she complains of a violent headache," Adam Beuvius, *Henrietta of Grestenfeld* (London: William Lane, 1787–88), 18. For an overview on the medicine in eighteenth-century fiction, see Roy Porter, "Lay Medical Knowledge in the Eighteenth Century: The Evidence of the *Gentleman's Magazine*," *Medical History* 29, No. 2 (1985)

90. James Beresford, *The Miseries of Human Life or the Groans of Samuel Sensitive and Timothy Testy* (Sixth edition, London: J. Ballantyne, 1807), 262.

91. Thomas Reid, *An Essay on the Nature and Cure of the Phthisis Pulmonalis* (London: T. Cadel, 1785), 21–22.

92. For a recent discussion of exposure as "contact between misplaced matter and flesh," see Gregg Mitman, Michelle Murphy, and Christopher Sellers (eds.), *Landscapes of Exposure: Knowledge and Illness in Modern Environments* (Chicago: University of Chicago Press, 2004).

93. *London's Medical Repository* (1814), 123.

94. John Cheshire, *A Treatise upon the Rheumatism* (London: C. Rivington, 1735), 10.

95. Thomas Sydenham, *The Entire Works of Dr. Thomas Sydenham* (London: Edward Cave, 1742), 283; Leonard Gillispie, *Advice to the Commanders and Officers of His Majesty's Fleet Serving in the West Indies* (London: J. Cuthell, 1798), 28; William Heberden, "On the Influence of Cold upon the Health of the Inhabitants of London," *Philosophical Transactions* 86 (1796): 279–84, 284; James Curry, *Popular Observations on Apparent Death from Drowning, Suffocation etc* (Northampton, 1792), 4; Bernardino Ramazzini, *Health Preserved* (Second Edition, London: John Whiston, 1750), 100, passim.

96. Michael O'Ryan, *Advice, in the Consumption of the Lungs* (Dublin: H. Fitzpatrick, 1798).

97. Henry Robertson, *A General View of the Natural History of the Atmosphere* (Edinburgh: Abernethy and Walker, 1808), 296.

98. Willan, *Reports on the Diseases in London*, 122. See also W.F. Bynum, "Cullen and the Study of Fevers in Britain, 1760–1820," in W.F. Bynum and V. Nutton (eds), *Theories of Fever from the Antiquity to the Enlightenment* (Medical Supplement No. 1, London: Wellcome Institute for the History of Medicine, 1981), 135–47, 142–43.

99. Joseph Chitty, *A Summary of the Office and Duties of Constables* (London: Shaw and Sons, 1837), 106.

100. William Falconer, *Remarks on the Influence of Climate, Situation, Nature of Country, Population, Nature of Food, and Way of Life, on the Disposition and Temper* (London: C. Dilly), 426.

101. *The Anti-Craftsman: Being an Answer to the Craftsman Extraordinary* (London: Brindley, 1729), 46; Country Gentleman, *A Narrative in Justification of Injured Innocence* (London: W. Webb, 1749); Emanuel Collins, *Lying Detected; Or, Some of the Most Frightful Untruths That Ever Alarmed the British Metropolis, Fairly Exposed* (London: E. Farley, 1758); Edward Bancroft, *Remarks on the Review of the Controversy between Great Britain and Her Colonies. In Which the Errors of Its Author Are Exposed* (London: T. Becket, 1769); William Cobbett, *The Republican Judge: Or the American Liberty of the Press, as Exhibited, Explained, and Exposed* (Second edition. London: J. Wright, 1798).

102. John Hemet, *Contradictions, or Who Would Have Thought It* (London: Earl and Hemet, 1799), 42–43.

103. *A New System of Practical Domestic Economy: Founded on Modern Discoveries and the Private Communications of Persons of Experience* (London: Colburn 1824), 131.

104. *Angeline*, vol. 1, 45.

105. Foundling Hospital (London, England). *An Account of the Hospital for the Maintenance and Education of Exposed and Deserted Young Children* (London: n.p. 1759). Chambers's *Encyclopeadia* defines "exposing" as the act of setting a thing to public view; a "sacrament is said to be exposed when it is shown in public, uncovered, on festival days." A house can have a good prospect, but it can be "exposed to all the four winds." "Exposing of children, a barbarous custom practiced by most of the ancients, whereby it was made capital to expose children, ordaining, at the same time, that such as were not in a condition to educate them should bring them to the magistrates in order to be brought up by the public expense." In gardening, "the aspect or situation of a garden, wall building, etc,, with respect to the sun, wind etc. east, west, north, south," in Ephraim Chambers, *Cyclopaedia; Or, An Universal Dictionary of Arts and Sciences* (Dublin: J. Chambers, 1780), vol. 2.

106. La Caze, quoted in Rey, "Vitalism, Disease and Society," 465.

107. Dorothy Holland and Andrew Kipnis, "Metaphors for Embarrassment and Stories of Exposure: The Not-So-Egocentric Self in American Culture," *Ethos* 22 (1994): 316–42, 330. See also Richard Shweder and Edmund Bourne, "Does the Concept of Person Vary Cross Culturally," in R. Shweder and R. LeVine (eds.), *Culture Theory* (Cambridge: Cambridge University Press, 1984).

2 Cursed by Comfort

1. Cheyne, *An Essay of Health and Long Life*, 6

2. Joseph Priestley, *Experiments and Observations on Different Kinds of Air* (3 vols. London: J. Johnson, 1777).

3. Woodruff D. Smith, "Complications of the Commonplace: Tea, Sugar, and Imperialism," *Journal of Interdisciplinary History* 23 (1992): 259–78.
4. See Mark Jackson, *Health and the Modern Home* (London: Routledge, 2008).
5. But see Corbin, *The Foul and the Fragrant*, and Margaret Jacob, "The Mental Landscapes of the Public Sphere," *Eighteenth Century Studies* (1994): 95–113.
6. Robin Evans, *The Fabrication of Virtue: English Prison Architecture, 1750–1840* (Cambridge: Cambridge University Press, 1982).
7. Colin Chisholm, "On the Statistical Pathology of Bristol and Clifton," *Edinburgh Medical Journal* 13 (1817): 265–93. On urban space, see Mark Jenner, "Underground, Overground: Pollution and Place in Urban History," *Journal of Urban History* 24 (1997): 97–110; and Christine Mesiner Rosen and Joel Arthur Tarr, "The Importance of an Urban Perspective in Environmental History," *Journal of Urban History* 20 (1994): 299–310.
8. Robert Jackson, *An Outline of the History and Cure of Fever, Endemic and Contagious* (Edinburgh: Mundell and Son, 1798), 110.
9. James Johnson, *Change of Air or Pursuit of Health* (London: S. Highley, 1839), 10.
10. Maxine Berg, *Luxury and Pleasure in Eighteenth Century Britain* (Oxford: Oxford University Press, 2005); John Sekora, *Luxury: The Concept in Western Thought, Eden to Smollett* (Baltimore and London: Johns Hopkins University Press, 1977).
11. Nathaniel Lancaster, *Public Virtue: Or, the Love of Our Country* (London: R. Dodsley, 1746), 25.
12. *The Idler by the Author of the Rambler* (London: J. Rivington, 1790), 51. "Surely nothing is more reproachful to a being endowed with reason, than to resign its powers to the influence of the air, and live in dependence [sic] on the weather and the wind." While it might have been uncontroversial that the seasons dictated epidemics and agriculture, "to call upon the Sun for peace and gaiety, or deprecate the Clouds lest sorrow should overwhelm us, is the cowardice of Idleness, and the idolatry of Folly."
13. Tobias George Smollett, *The Expedition of Humphry Clinker* (London: T. Johnston, 1771), 161. See Michael McKeon, "Aestheticising the Critique of Luxury," in Berg and Eger (eds.), *Luxury in the Eighteenth Century,* 57–70.
14. Tobias Smollett, *The Expedition of Humphry Clinker* (London: Harrison and Co., 1785) vol. 2, 5.
15. Roy Porter, "Diseases of Civilization," in W.F. Bynum and Roy Porter (eds.), *Companion Encyclopedia of the History of Medicine* (London: Routledge, 1993), 585–600. Roy Porter, "Nervousness, Eighteenth and Nineteenth Century Style: From Luxury to Labour, in Marijke Gijswijt-Hofstra and Roy Porter (eds.), *Cultures of Neurasthenia*

from Beard to the First World War (Amsterdam: Rodopi, 2001), 31–47; Vladimir Janković "Arcadian Instincts: A Geography of Truth in Georgian England," in Miles Ogborn and Charles W.J. Withers, *Georgian Geographies: Essays on Space, Place, and Landscape in the Eighteenth Century* (Manchester: Manchester University Press, 2004), 174–91.

16. Richard Price, "Observation on the Difference between the Duration of Human Life in Towns and Country Parishes and Villages," *Philosophical Transactions* 65 (1683–1775): 424–445, 428.

17. *Hippocrates's Treatise on the Preservation of Health*, 4–5.

18. Percival Stockdale, *Three Discourses: Two Against Luxury and Dissipation. One on Universal Benevolence* (London: W. Flexney, 1773), 8.

19. Falconer, *Remarks on the Influence of Climate*, 508, 4.

20. James Stephen Taylor, "Philanthropy and Empire: Jonas Hanway and the Infant Poor of London," *Eighteenth-Century Studies* 12, No. 3 (1979): 285–305

21. Jonas Hanway, *Serious Considerations on the Salutary Design of the Act of Parliament* (London: John Rivington, [1762]), 48–49.

22. Hanway, *Serious Considerations*, 52; idem, *Midnight the Signal* (London: Dodsley, 1779), 78.

23. *Boerhaave's Aphorisms: Concerning the Knowledge and Cure of Diseases* (Third Edition, London: W. Innys, 1755), 247; medical benefits of plowing are mentioned in Robert Brookes, *The General Practice of Physic* (London: J. Newbery, 1754), vol. 1, 302.

24. Daniel Cox, *Observations on the Intermitting Pulse, as Prognosticating, in Acute Diseases* (London: A. Millar, 1758), 96; Richard Drake, *An Essay on the Nature and Manner of Treating the Gout* (London: Author, 1758), 77; *The Fountain of Knowledge, or, British Legacy* (London: Bailey, 1760), 82, 97

25. William Falconer, *An Essay on the Preservation of the Health of Persons Employed in Agriculture* (London: R. Cruttwell, 1789). On the cultures of urban and rural space, see Raymond Williams, *The Country and the City* (London: Chatto and Windus, 1973).

26. William Falconer, *Observations Respecting the Pulse* (London: T. Cadell, 1796).

27. Michael Dorn, "Climate, Alcohol, and the American Body Politic: The Medical and Moral Geographies of Daniel Drake (1785–1852), PhD dissertation, University of Kentucky, 2003; Felix Driver, "Moral Geographies," *Transactions of the Institute of British Geographers* 25 (1988): 333–46; a similar argument is made in Geoffrey Harding, *Opiate Addiction, Morality and Medicine: From Moral Weakness to Pathological Disease* (Basingstoke: MacMillan, 1986).

28. James Gregory, *A Dissertation on the Influence of Change of Climate in Curing Diseases* (Philadelphia: T. Dobson, 1815), 29.

29. Gregory, *Dissertation*, 30.

30. James Makittrick Adair, *Medical Cautions, for the Consideration of Invalids; those Especially who Resort to Bath: Containing Essays on Fashionable Diseases* (Bath: R. Cruttwell, 1786), 12.

31. Adair, *Medical Cautions*, 43.

32. Adair, *Medical Cautions*, 43.

33. Smollett, *Humphry Clinker*, 85–87.

34. Adair, *Medical Cautions*, 35. William Buchan advised that closed atmospheres weakened the child's constitution, which he compared to greenhouse plants not destined for "the strength, vigour and magnitude" of their outdoor counterparts. William Buchan, *Domestic Medicine* (London: W. Strachan, 1772), 39.

35. Tim Meldrum, "Domestic Service, Privacy and the Eighteenth-Century Metropolitan Household," *Urban History* 26 (1999): 27–39, 38.

36. Adair, *Medical Cautions*, 44.

37. *The Times*, May 10 1802, 2. A good-sized room (twenty by thirteen feet wide, and ten feet high) would contain two hundred pounds of air. A single pound of charcoal (roughly equivalent to ten candles) would burn a quantity of oxygen contained in fourteen pounds of atmospheric air. See Charles Tomlison, *Rudimentary Treatise on Warming and Ventilation* (London: John Weale, 1850), 52.

38. Jan Dalley, *The Black Hole: Money Myth, and Indian Empire* (London: Penguin, 2006).

39. Willich expanded the content of his *Lectures* in *The Domestic Encyclopaedia; or, A dictionary of Facts, and Useful Knowledge* (4 vols. London: Murray and Highely, 1802); the American edition from 1803–4 was issued in five volumes. Willich was favorably quoted by Alexander Hunter, George Gregory, and John Sinclair, among others.

40. Willich, *Lectures on Diet*, 181ff. On the chemistry of candle burning, see Bryan Cornwell, *The Domestic Physician; or, Guardian of Health* (London: J. Murray, 1784), 64–65.

41. James Anderson, *A Practical Treatise on Chimneys* (Edinburgh, C. Elliot, 1776).

42. Robert Bath, *An Essay on the Medical Character* (5th ed., London: C. Laidler, [1790?]) 137–38.

43. Jan Golinski, *British Weather and the Climate of Enlightenment* (Chicago: University of Chicago Press, 2007).

44. Thomas Beddoes, Thomas Beddoes, *Essay on the Causes, Early Signs, and Prevention of Pulmonary Consumption for the Use of Parents and Preceptors* (London: Longman and Rees, 1799), 133.

45. Thomas Beddoes, *A Lecture Introductory to a Course of Popular Instruction on the Constitution and Management of the Human Body,* (Bristol: N. Biggs, 1797), 27.

46. George Adams, *Lectures on Natural and Experimental Philosophy,* (5 vols. London: R. Hidmarsh, 1794), vol. 1, 70. Confined nurseries

were claimed to relax children's fibers "and rendered delicate and liable to cold [...] even the cautions of affluence, and the arts introduced by luxury to render the habitations of the wealthy impervious to cold, are extremely prejudicial." *The Compleat Family Physician* (Newcastle upon Tyne: printed by Matthew Brown, 1800–1801), 23.

47. John Whitehead, *The Life of the Rev. John Wesley, M.A. Some Time Fellow of Lincoln-College, Oxford* (London, 1793–96), 50, 1.

48. James Graham, *A New and Curious Treatise of the Nature and Effects of Simple Earth, Water, and Air, etc.* (London: Richardson, 1793), 5. For Van Swieten's reports on these, see *Annual Register for the Year 1765* (4th ed., London: J. Dodsley, 1784), 108. On Graham, see Barbara Brandon Schnorrenberg, "A True Relation of the Life and Career of James Graham, 1745–1794," *Eighteenth-Century Life* 15 (1991): 58–75, Roy Porter, *Quacks: Fakers and Charlatans in Medicine* (London: Tempus, 1989).

49. Joshua White, *Letters on England* (Philadelphia, for the author, 1816), vol. 1, 22. See also John Cornforth, *London Interiors: From the Archives of Country Life* (London: Aurum Press, 2000); idem, *English Interiors, 1790–1848: The Quest for Comfort* (London: Barrie and Jenkins, 1978); Alan and Ann Gore, *The History of English Interiors* (Oxford: Phaidon, 1991); Elizabeth Burton, *The Georgians at Home 1714–1830* (London: Longmans, 1967), 109–153.

50. Hans-Joachim Voth, "Time and Work in Eighteenth-Century London," *The Journal of Economic History* 58 (1998): 29–58.

51. Richard Brinsley Sheridan, *The School for Scandal* (London: E. Powell, 1798), 14.

52. Duden, *The Woman Beneath the Skin*, 141.

53. William Godwin, *Things as They Are* (London: G.G. and J. Robinson, 1796), vol. 3, 22.

54. Thomas Beddoes, *Hygiea* (London: J. Mills, 1802), 174; James Stephen, *The Dangers of the Country* (London: J. Butterworth and J. Hatchard, 1807), 145.

55. Thomas Beddoes, *A Guide for Self-Preservation* (Bristol: Bulgin and Rosser, 1794), iv.

56. John E. Crowley, *The Invention of Comfort: Sensibilities and Design in Early Modern Britain and Early America* (Baltimore and London: Johns Hopkins University Press, 2001), 194–200.

57. Beresford, *Miseries*, 13.

58. Beresford, *Miseries*, 215, 223. On travails associate with entertainment: 5 (Samuel) "After the play, on a raw wet night, with a party of ladies—fretting and freezing in the outer lobbies, and at the street-doors of the theatre, among chair-men, barrow-women, yelling link-boys, and other human refuse, in endless attempts to find out your servant, or carriage, which, when found at last, cannot be drawn up nearer than a furlong from the door."

59. Sinclair, *The Code of Health and Longevity*, Vol. 1, 737–747, 744.
60. *A New System*, 70.
61. *A New System of Practical Domestic Economy*, 131.
62. Virginia Sarah Smith, *Clean: A History of Personal Hygiene and Purity* (Oxford: Oxford University Press, 2007), 217
63. On "shutting in," see A. Roger Ekirch, *At Day's Close: Night in Times Past* (New York: W. W. Norton and Company, 2005), 14–15, 93, 270,
64. *Encyclopedia Britannica* (Edinburgh: A. Bell and C. Macfarquhar, 1771), vol. 1, 1: "aerophobia, among physicians signifies the dread of air," John Ash, *The New and Complete Dictionary of the English Language* (London: Edward and Charles Dilly, 1775). *Aerophobi*, according to Caelius Aurelianus, were those afraid of either a "lucid" or "obscure" air, as in George Motherby, *A New Medical Dictionary* (London: J. Johnson, 1775). See also Giambattista Morgagni, *The Seats and Causes of Diseases Investigated by Anatomy* (London: A. Millar, 1769), 172. On aerophobia and hydrophobia, see John Aitken, *Elements of the Theory and Practice of Physic and Surgery* (London, n.p., 1783), 534. For the context of Sauvage's nosology, see Elizabeth A. Williams, "Hysteria and the Court Physician in Enlightenment France," *Eighteenth-Century Studies* 35(2002) 247–255.
65. Benjamin Franklin, *Observations on Smoky Chimneys* (London: John Debrett, 1787), 30.
66. Bisset, *An Essay on the Medical Constitution of Great Britain*, 6.
67. Jan Golinski argues that the emphasis on rapidity of change followed the widespread domestic use of barometers, an instrument that records short-term fluctuations of weather, rather than slow climate patterns; Golinski, "The Human Barometer," 2. See also Jones and Brockliss, *The Medical World of Early Modern France*, 463.
68. *A New System of Practical Domestic Economy*, 141.
69. Mackenzie, *The History of Health*, 369–70.
70. Martha Bradley, *The British Housewife: or, the Cook, Housekeeper's, and Gardiner's Companion* (London, [1760?]).
71. *Domestic Economy; or, Complete System of English Housekeeping* (London: J. Creswick 1794), 333.
72. *An Enquiry into the Causes of the Present Epidemical Diseases* (London: F. Fayram, 1729), 2.
73. Quoted in John Fowler and John Cornforth, *English Decoration in the 18th Century* (London: Barrie and Jenkins, 1974), 225.
74. *British Foreign and Medical Review* 15 (1843): 129.
75. Peter Kalm, *Visit to England* (1748) quoted in Burton, *The Geogians at Home 1714–1830*, 118–119.
76. John Wood, *A Series of Plans for Cottages or Habitations of the Labourer* (London: I. and J Taylor, 1792), 2.

77. Adair, *Medical Cautions*, 36. Fevers were known to begin when cold and moist weather succeeded a long period of dry and sultry weather, according to Bisset, *An Essay on the Medical Constitution of Great Britain*, 6.

78. Mark Girouard, *Life in the English Country House: A Social and Architectural History* (New Haven and London: Yale University Press, 1978), 194.

79. Lawrence Stone, "The Private and the Public in the Stately Homes in England, 1500–1990," *Social Research* 58 (1991): 227–57, 256. Carole Shammas, "The Domestic Environment in Early Modern England and America," *Journal of Social History* 14 (1980): 4–24; Frank E. Brown, "Continuity and Change in the Urban House: Developments in Domestic Space Organisation in Seventeenth-Century London," *Comparative Studies in Society and History* 28 (1986): 558–590. On innovation in domestic comfort across Europe, see Rafaella Sarti, *Europe at Home: Family and Material Culture 1500–1800* (New Haven: Yale University Press, 2002), 96ff.

80. Bridget Hill, *Servants: English Domestics in the Eighteenth Century* (Oxford: Clarendon Press, 1996), 39–41; J. Jean Hecht, *The Domestic Servant Class in Eighteenth-Century England* (Routledge & Kegan Paul, London, 1956).

81. James Makittrick Adair, *Essay on Fashionable Diseases* (London: T.P. Bateman, 1790), 70.

82. See Dorothy Marshall, "Review G.E. and K.R. Fussel's *The English Countrywoman*," *Economic History Review* 7 (1954); 109–10.

83. Walter Bernan, *On the History and Art of Warming and Ventilating Rooms and Buildings* (London: George Bell, 1845), 90, iv.

84. Robert Thornton, *Medical Extracts* (3 vols. London: J. Johnson, 1796), vol. 2, preface.

85. On living conditions of European peasantry, see Jerome Blum, *The End of the Old Order in Rural Europe* (Princeton: Princeton University Press, 1978), 178–183. On timber hardening, see Peter Brimblecombe, "Interest in Air Pollution among Early Fellows of the Royal Society," *Notes and Records of the Royal Society* 32 (1978): 123–129; Edmund Newell, "Atmospheric Pollution and the British Industry, 1690–1920," *Technology and Culture* 38 (1997): 655–689.

86. *Daily Gazetteer*, January 12, 1740.

87. *General Evening Post*, January 17, 1740.

88. Hector Gavin, *Sanitary Ramblings* (London: John Churchill, 1848), 70.

89. Hanway, *The Citizen's Monitor* (London: Dodsley, 1780), 145.

90. *The Economist and General Adviser* (London: Knight and Lacey, 1825), 151.

91. Adair, *Medical Cautions*, 53.

92. Coleman, "Health and Hygiene in the Encyclopaedia," 406.

3 Artificial Airs

1. *The Universal Family Physician and Surgeon* (Blackburn: Hemingway and Nuttall, 1798), 106.
2. Owsei Temkin, "An Historical Analysis of the Concept of Infection," in G. Boas et al. (eds.), *Studies in Intellectual History* (Baltimore: Johns Hopkins University Press, 1953), 123–47.
3. This compilation is based on Joel Pinney, *An Exposure of the Causes of the Present Deteriorated Health* (London: Longman, 1830); John Roberton, *General Remarks on the Health of English Manufacturers* (London: James Ridgway, 1831); Francis Lloyd, *Practical Remarks on the Warming, Ventilation, and Humidity of Rooms* (London: George Cox, 1854).
4. William Cullen, *Synopsis nosologiæ methodicæ* (Edinburgh: W. Creech, 1780), 183; Aitken, *Elements of the Theory and Practice of Physic and Surgery,* vol. 1, 496.
5. Joseph Townsend, *A Guide to Health* (London: Cox, 1795–96), 81.
6. For a recent discussion, see George Crile (Amy Rowland, ed.), *Diseases Peculiar to Civilized Man: Clinical Management and Surgical Treatment* (New York: MacMillan, 1934); C.F. Wooley, "Where Are the Diseases of Yesteryear? DaCosta's Syndrome, Soldiers Heart, the Effort Syndrome, Neurocirculatory Asthenia—and the Mitral Valve Prolapse Syndrome?" *Circulation* 53 (1976): 749–751; G. E. Berrios, "Feelings of Fatigue and Psychopathology: A Conceptual History," *Comprehensive Psychiatry* 31 (2) (1990): 140–151.
7. E. Shorter, "Chronic Fatigue in Historical Perspective," *Ciba Foundation Symposium* 173 (1993): 16–22.
8. Thomas Withers, *Observations on the Abuse of Medicine* (London: J. Johnson, 1775), 130ff.
9. See Thomas Withers, *Observations on Chronic Weakness* (York: A. Ward, 1777).
10. Withers, *Observations,* 31.
11. John Fitchen, "The Problem of Ventilation through the Ages," *Technology and Culture* 22 (1981): 485–511; Bernard Nagengast, *Heat and Cold: Mastering the Great Indoors* (Atlanta: ASHRAE, 1994); G.L. Colice, "Historical Perspective on the Development of Mechanical Ventilation," in M.J. Tobin (ed.), *Principles and Practice of Mechanical Ventilation* (New York: McGraw-Hill, 1994). On air in hospitals, see Jeanne Kisacky, "Restructuring Isolation: Hospital Architecture, Medicine, and Disease Prevention," *Bulletin for the History of Medicine* 79 (2005): 1–49; Christine Stevenson, *Medicine and Magnificence: British Hospital and Asylum Architecture, 1660–1815* (New Haven: Yale University Press, 1999); Fielding H. Garrison, "The History of Heating, Ventilation, and Lighting," in *Bulletin of the New York Academy of Medicine* 3 (1927): 57–67, 26–41.

12. J. Durno, *A Description of a New-Invented Stove-Grate, Shewing Its Uses and Advantages over All Others* (London: J. Towers, 1753), 2.

13. Samuel Sutton, *An Historical Account of a New Method for Extracting the Foul Air out of Ships* (London: J. Noon, 1745), 2–3. Another early experiment is described in J.T. Desaguliers, "An Account of an Instrument or Machine for Changing the Air of the Room of Sick People in a Little Time, by Either Drawing out the Foul Air, or Forcing in Fresh Air; or Doing Both Successively, without Opening Doors or Windows," *Philosophical Transactions of the Royal Society of London* 39 (1735–36): 41–43.

14. Stephen Hales, *A Description of Ventilators Whereby Great Quantities of Fresh Air May with Ease Be Conveyed into Mines, Gaols, Hospitals, Work-houses, and Ships, in Exchange for Their Noxious Air* (London: W. Innys, 1743), 39. Fraser Harris, "The Pioneer in the Hygiene of Ventilation," *Scientific Monthly* 3 (1916): 440–454; Peter Collinson, "Some Account of the Life of the Late Excellent and Eminent Stephen Hales," *Annual Register* 7 (1764): 42–49; Jocelyn Thorpe, "Stephen Hales, 1677–1761," *Notes and Records of the Royal Society of London* 3 (1940): 53–63.

15. Stevenson, *Medicine and Magnificence,* 170.

16. Jeremy Bentham, *Management of the Poor* (Dublin: James Moore, 1796), 245.

17. Enid Gauldie, *Cruel Habitations: A History of Working-Class Housing 1780–1918* (London: Routledge, 1974), 93.

18. John Whitehurst and Robert Willan, *Observations on the Ventilation of Room* (London: W. Bent, 1794).

19. Hanway, *Serious Considerations,* 54.

20. Lynch, *A Guide to Health,* 103.

21. Temkin, "The Concept of Infection," 139–40.

22. Jonas Hanway, *The Citizen's Monitor: Shewing the Necessity of a Salutary Police* (London: Dodsley, 1780), 131. On Hanway, see James Stephen Taylor, *Jonas Hanway: Founder of the Marine Society: Charity and Policy in Eighteenth Century Britain* (London: Scolar Press, 1985).

23. John Carter, *The Builder's Magazine* (London: Newberry, 1774–8), 297. John Whitehurst (or Robert Willan, in the posthumous edition of his work on air ducts) reminisced, "[A]lthough the salutary tendency of [my] plan must be obvious when fully considered, there still remains a prejudice against it in the minds of the multitude. They obstinately maintain that the same injuries are to be expected from air admitted in the manner, as from the cold streams of it which usually flow into a room through the crevices of the door or window." Whitehurst and Willan, *Observations on the Ventilation of Rooms,* 19.

24. Mackenzie, *The History of Health and the Art of Preserving It,* 369.

25. Beresford's chapter on miseries of London gives this morose description of home moisture: "The meridian midnight of a thick London

fog—leaving you no method of distinguishing between the pavement, and the middle of the street; much less between one street and another—the 'palpable obscure' pursuing you into your parlour, and bed-chamber, till you can neither see, speak, nor breath." Beresford, *Miseries of Human Life*, 78. Herman Mutthesius remarked, "Humidity is the most marked of all England's climatic features: it accounts for the frequent fogs, for the distances in which all is veiled in a light mist, for the luxuriant green of the plant life [...] for that permanent chilly feeling that so often causes visitors to catch cold. In enclosed spaces it induces the fusty, musty smell that affects any room that is not thoroughly aired every day." Hermann Muthesius, *The English House* (Ed. Dennis Sharp. London: Crossby Lockwood Staples (1904), 1979), 67.

26. Samuel Madden, *Memoirs of the Twentieth Century* (London: Osborn and Longman, 1733), 70.

27. Mackenzie, *The History of Health*, 368–69.

28. Buchan, *Domestic Medicine*, 52. On this point, see Christopher Lawrence, "William Buchan: Medicine Laid Open," *Medical History* 19 (1975): 20–35.

29. Emily Cockayne, *Hubbub: Filth, Noise, and Stench in England, 1600–1770* (New Haven: Yale, 2007).

30. William Chambers, *A Treatise on Civil Architecture* (London: J. Haberkorn, 1759), i.

31. Chambers, *Treatise*, 70.

32. "On Architecture," in John Aheron, *A General Treatise of Architecture* (5 vols., Dublin: John Butler, 1754), vol. 2, 81. See also *Encyclopedia Brittanica* (2nd ed., 10 vols., Edinburgh: J. Balfour), vol. 1, 615: "The first consideration with regard to windows is their size which varies according to the climate."

33. Robert Clavering, *An Essay on Construction and Building of Chimneys* (London: I. Taylor, 1799), 57 ff.

34. W.R. Ward, "The Administration of the Window and Assessed Taxes, 1696–1798," *The English Historical Review* 67 (1952): 522–542; S. Dowell, *A History of Taxation and Taxes in England* (London, 1884), vol. 2., 187–88.

35. Timothy Silence, *The Foundling Hospital for Wit* (London: W. Webb, 1747), 47: "Jove said, Let there be Light—and/It instant was; and freely given/To every Creature under heaven/Says P—m 'I'll not have it so;/'Darkness much better suits my Views,/let Darkness o'er the Land diffuse./Henceforth I will that all shall pay,/For every Light, by Night or Day.' P—m's defense is: 'Open your Door, or go without it, and Air is free, you cannot doubt it: Nay, in your Windows there are Cracks, Where Air finds Passage—without Tax."

36. Anthony Fothergill, *An Inquiry into the Suspension of Vital Action* (Bath: S. Hazard, 1795), 75. Discussion on the pathological effects of light privation is in John Howard, *An Account of the Present State*

of the Prisons (London: Society lately instituted for giving effect to His Majesty's proclamation against vice and immorality, 1789); Thornton's *Medical Extracts,* vol. 3, 434.

37. Ferriar, *Medical Histories and Reflections,* 246.

38. For example, Robert James, *A Dissertation on Fevers and Inflammatory Distempers* (London: Francis Newberry, 1770), 66.

39. James Lind, *An Essay on the Most Effectual Means of Preserving the Health of Seamen in the Royal Navy* (London: J. Murray, 1778), 334–35.

40. Ferriar, *Medical Histories and Reflections,* vol. 2, 187;

41. Adair, quoted in Sinclair, *The Code of Health and Longevity,* vol. 1, 735.

42. Edmund Gillingwater, *An Essay on Parish Work-Houses* (Bury St. Edmund's: J. Rackham, 1786), 37. See also Kathryn Morrison, *The Workhouse: A Study of Poor Law Buildings in England* (Swindon: Royal Commission on the Historical Documents of England, 1999), 20 ff.

43. *The Child's Physician* (London: P. Boyle, 1795), 125.

44. William Falconer, *An Account of the Use, Application, and Success of the Bath Waters, in Rheumatic Cases* (Bath: W. Meyler, 1795), 20.

45. Townsend, *Guide to Health,* 109.

46. Benjamin Moseley, *A Treatise on Tropical Diseases* (London: G.G. and J. Robinson, 1795), 168.

47. Beddoes, *Manual of Health,* 164; on the benefits of open air in typhus, see David Paterson, *A Treatise on Scurvy* (Edinburgh: Manners and Miller, 1795), 57.

48. Hentie J. Louw and Robert Crayford, "A Construction History of the Sash Window: Part 1," *Architectural History* 41 (1998): 82–130; and idem, "A Construction History of the Sash Window: Part 2," *Architectural History* 42 (1999): 173–239.

49. Adair, *An Essay on Regimen,* 98.

50. Low and Crayford, "A Construction," 185. Double-glazing appeared during the 1640s as a weatherproofing technique with adverse effects to daylight.

51. Jonathan Swift, *Directions to Servants in General* (London: R. Dodsley, 1745), 78–79.

52. Hanway, *Serious Considerations,* 51.

53. In France, Tissot wrote: "Not to renew the air of one's room, is to live in the impurities of the preceding day, and yet what hard student is there who thinks of letting fresh air into his chamber every day?" Samuel Auguste David Tissot, *A Treatise on the Diseases Incident to Literary and Sedentary Persons* (Edinburgh: Donaldson, 1772), 51.

54. Hanway, *Citizen's Monitor,* 131.

55. Thomas Tidd, *Considerations on the Use and Properties of the Aeolus a New Invented Portable Machine for Exchanging and Refreshing the Air of Rooms etc.* (London: J. Reeves, 1755). For another example

of portable ventilation, see Benjamin Martin, *The Young Gentleman and Lady's Philosophy*, (Third edition, London, 1781), vol. 1, 408.

56. Tidd, *Considerations*, 4.

57. Tidd, *Considerations*, 7. Tidd refers to Huxham's observation that "a close, narrow, stifling room is exceedingly incommodious to any person sick of a fever, but more to those who are ill of peripneumoni." The reputation of Tidd's invention went well into the nineteenth century, as witnessed in Alexander Jamieson's complimentary description in his *A Dictionary of Mechanical Sciences, Arts, Manufactures, and Miscellaneous Knowledge* (London: Henry Fisher, 1829), 12: "Aeolus, in mechanics, a small portable machine for refreshing and changing the air in rooms, adapted in its dimensions to supply the place of a square of glass in a sash window; and executed in so small a compass as to project but a little way from the sash; and in so neat a manner as to be an elegant ornament to the place where it is fixed. It works without at least noise, requires not attendance, and occasions neither trouble not expense to keep it in order. It throws in only such as quantity of air, as is agreeable, and leave off working of itself wherever the door or window is opened." The "agreeability" of modern interiors was cited as the most obvious reason in opting for nontraditional technologies, as was the case with Count Rumford's versions of American stove, commercially promoted in England by James Sharp in *An Account of the Principle and Effects of the Air Stove-Grates* (Tenth edition. [London], [1785]). See Priscilla J. Brewer, *From Fireplace to Cookstove: Technology and the Domestic Ideal in America* (New York: Syracuse University Press, 2000), 15–37.

58. Sir Gilbert Blane, *Observations on the Diseases of Seamen* (London: Joseph Cooper, 1799).

59. There are many examples of medical writing, especially in health manuals, in which the practitioners advise on the methods of airing the rooms; e.g., Adair, *Medical Cautions*, 47, 94; Sinclair, *Code of Health and Longevity*, 736, quoting Adair.

60. J.C. Loudon, *Remarks on the Construction of Hothouses* (London: J. Talor, 1817).

61. Anderson, *A Practical Treatise on Chimneys*, 44.

62. Anderson, *A Practical Treatise*, 44.

63. Loudon, *Remarks*, 69.

64. Robert Bruegmann, "Central Heating and Forced Ventilation: Origins and Effects on Architectural Design," *The Journal of the Society of Architectural Historians* 37 (1978): 143–60.

65. Jean-Frédéric de Chabannes, Marquis de Curton, *Explanation of a New Method for Warming and Purifying the Air in Private Houses and Public Buildings* (London: Shulze and Dean, 1815), 6, 11.

66. Engineer, *The Theory and Practice of Warming and Ventilating Public Buildings, Dwelling Houses, and Conservatories* (London: Thomas and George Underwood, 1825), preface; a detailed account is given

in Whitehurst and WIllan, *Observations on the Ventilation of Rooms* (1794).

67. *The Times*, Friday, Jun 13, 1794, 1.

68. William White, *By His Majesty's Royal Letters Patent. An Air Machine.* [London], [1790?], 5; the machine also advertised in the *Times*, Thursday, Jun 25, 1789, 1.

69. *Times*, Thursday, Nov 19, 1789, 2.

70. Sinclair, *Code of Health and Longevity*, 736 quotes Adair's *Medical Cautions*, 47, 94.

71. Georges Teyssot, "Habits/Habitus/Habitat," *Present and Futures. Architecture in Cities* (Barcelona: CCCB, 1996), 1. Jean-Frédéric Chabbanes, *Prospectus d'un projet pour la construction de nouvelles maisons dont tous les calculs et les détails procureront une très grande économie et beaucoup de jouissance* (Paris: Desenne 1803).

72. Anthony Meyler, *Dissertatio medica inauguralis, de melancholia* (Edinburgh: C. Stewart, 1803).

73. *The Stranger in Liverpool: Or, an historical and Descriptive View of the Town of Liverpool and Its Environs* (7th ed., Liverpool: T. Kaye, 1823), 164.

74. Meyler, *Observations on Ventilation*, 194.

75. Meyler, *Observations on Ventilation*, 3.

76. Meyler, *Observations on Ventilation*, 2–3.

77. Meyler, *Observations on Ventilation*, 37.

78. Meyler, *Observations on Ventilation*, 37.

79. Meyler, *Observations on Ventilation*, 40.

80. This was commonplace by the time Gavin wrote, "[T]he air which is breathed within the dwellings of the poor is often most insufferably offensive *to strangers* [...] In numerous instances *I found* the air in the rooms of the poor [...] so saturated with putrescent exhalations, that to breathe it was to inhale a dangerous, perhaps fatal, poison." Gavin, *Sanitary Ramblings* (1848), 69.

81. Adair, *An Essay on Regimen*, 98.

82. "The Art of Procuring Pleasant Dreams," in *The Cabinet* (Edinburgh: R. Morison, 1797), 87–89.

83. Meyler, *Observations*, 43.

84. John Styles, "Product Innovation in Early Modern London," *Past and Present* 168 (2000): 124–69.

85. The issue of removability featured prominently in the public health movement; see Christopher Hamlin, *Public Health and Social Justice in the Age of Chadwick* (Cambridge: Cambridge University Press, 1998), 226. Ventilation attracted considerable attention in the period in publications such as Mr. J.B. Davis, *A Popular Manual of the Art of Preserving Health* (London: Whittaker and Co., 1836); Charles James Richardson, *A Popular Treatise on the Warming and Ventilation of Buildings* (London: John Weale, Architectural Library, 1837); William Harley, *The Harleian Dairy System; Also, a New and*

Improved Mode of Ventilating Stables (London: J. Ridgway, 1829); Robert Stuart Miekleham, *The Theory and Practice of Warming and Ventilating Public Buildings, Dwelling-houses, and Conservatories* (London: T. and G. Underwood, 1825); Thomas Tredgold, *Principles of Warming and Ventilating Public Buildings, Dwelling-houses, Manufactories, Hospitals, Hot-houses, Conservatories* (London: Taylor, 1824), etc.

86. Meyler, *Observations on Ventilation*, 44.
87. For a more cautious approach, see Stephen Mosley, "Fresh Air and Foul: The Role of the Open Fireplace in Ventilating the British Home, 1838–1910," *Planning Perspectives* 18 (2003): 1–21.

4 Intimate Climates

1. Sutherland, *Attempts to Revive Antient Medical Doctrines*, 147.
2. Drew Leder, *The Absent Body* (Chicago: Chicago University Press, 1990).
3. Ruth O'Brien, Esther C. Peterson, and Ruby K. Worner, *Bibliography on the Relation of Clothing to Health* (Washington, DC: U.S. Dept of Agriculture, 1929); S. Levitt, & J. Tozer, *Fabric of Society: A Century of People and their Clothes, 1770–1870* (Carno: Laura Ashley, 1983); Beverly Lemire, *Fashion's Favourite: The Cotton Trade and the Consumer in Britain*, 1660–1800 (Oxford: Oxford University Press, 1991); Amy de la Haye and Elizabeth Wilson, *Defining Dress: Dress as Object Meaning and Identity* (Manchester: Manchester University Press, 1999).
4. Roy Porter, *Doctor of Society: Thomas Beddoes and the Sick Trade in Late Enlightenment England* (London: Routledge, 1992), 107.
5. Forster, *A Treatise on the Causes of Most Diseases*, 354.
6. John Chandler, *A Treatise of the Disease Called a Cold* (London: A. Millar, 1761), 82.
7. Forster, *Treatise*, 359. Procatarctic cause is described as that which sets the predisposing cause in motion.
8. Robert Hooper, *A Compendious Medical Dictionary* (London: Murray and Highley, 1798), 226.
9. The same procatarctic causes produce different effects depending on the individual predisposition (if two different "habits" are exposed to cold, plethoric, and dropsical, the first develops a pleurisy, the other stays healthy), *Dr. Boerhaave's Academical Lectures on the Theory of Physic*, vol. 6, 184; vol. 5: 381.
10. John Ball, *A Treatise of Fevers* (London: H. Cock, 1758), 25, mentions "the common or external procatarctic causes of Intermittent Fevers." Similar meaning is found in Samuel Chapman, *An Essay on the Venereal Gleet* (London: A. M'Culloh, 1751), 4; John Huxham in *An Essay on Fever* (London: J. Hinton, 1775), 18, writes of the common procatarctic cause of agues as "a moist foggy atmosphere

exhaling form swampy, morass soil, or from the continuance of cold, tainy thick weather." "The instability of English climate" is a procatarctic cause for Philip Stern, *Medical Advice to the Consumptive and Asthmatic People of England* (12th ed., London: J. Almon, 1771). The procatarctic cause could have mental origins, as explained in *Phthisiologia* (London: Thomas Boozey, 1798), lix: "therefore, trightly has Morton ranked amongst the procatarctic causes, affliction of the mind, initense attention and thought."

11. Santorio, *Medicina Statica: Being the Aphorisms of Sanctorius* (London: William Newton, 1728), 330, 329. Later works on the subject include John Lining, "Extracts of Two Letters from John Lining, Physician at Charles-Town in South Carolina to James Jurin Giving an Account of Static Experiments Made Several Times a Day upon Himself for One Whole Year," *Philosophical Transactions of the Royal Society* 42 (1746): 491–509; Bryan Robinson, *A Dissertation on the Food and Discharges of Human Bodies* (Dublin: S. Powell, 1747). See F.M. Valadez, "James Keill of Northampton, Physician, Anatomist and Physiologist," *Medical History* 15 (1971): 317–36.

12. Peter C. Baldwin, "How Night Air Became Good Air, 1776–1930," *Environmental History* 8 (2007): 412–29.

13. Issues of this nature were discussed at length in, among other works, Hugh Smythson, *The Compleat Family Physician; or, Universal Medical Repository* (London: Harrison, 1785), 198–99. Cornwell, *The Domestic Physician*, 302. Bath, *An Essay on the Medical Character*, 114; George Cheyne, *An Essay of the True Nature and Due Method of Treating the Gout* (London: G. Strahan, 1722), 104; Timothy Bennet, *An Essay on the Gout* (London: Richard Ford, 1734), 90; Brooks, *The General Practice of Physic*, 296, on sleeping in wet beds and clothing; for advice to travelers on how to avoid damp linen in inns, see Buchan, *Domestic Medicine*, 159–60; on accommodation on board of the Royal Navy vessels, see James Lind, *An Essay on the Most Effectual Means of Preserving the Health of Seamen in the Royal Navy*, 23. William Heberden suspected the fear of dampness to be overblown in his *Acta Collectanea Medica*, quoted in Hugo Owen, *Dissertatio medica, inauguralis, de contagione* (Edinburgh, Balfour and Smellie, 1783), 9–10; on dampness, see also Robert Harrington, *A Philosophical and Experimental Enquiry into the First and General Principles of Animal and Vegetable life* (London, T. Cadell, 1781), 144; on wet linen, see Willich's *Domestic Encyclopaedia*, 272.

14. Shaw, *A New Practice of Physic*, 323; see also Cheshire, *A Treatise Upon the Rheumatism*, 10, where the author identifies obstructed perspiration with "taking cold."

15. "Whatever disorders are primarily owing to the action of cold air, whether wet or dry, or of other cold and wet substances, may properly fall under this general denomination of a Cold, and such action may be called its procatarctic, or antecedent cause." The remote cause was

the obstruction of perspiration. Chandler, *Treatise*, 27. E. Bullman mentions headaches, fevers, and agues as caused by moisture, see his *The Family Physician* (London: for the author, 1789), 39. John Astruct connected soggy weather with menstrual disruptions in *A Treatise on All the Diseases Incident to Women* (London: T. Cooper, 1743), 79.

16. Thomas Apperley, *Observations in Physick* (London: W. Innys and J. Leake, 1731), 74.

17. William Alexander, *Experimental Essays on the Following Subjects: I. On the External Application of Antiseptics in Putrid Diseases. II. On the Doses and Effects of Medicines. III. On Diuretics and Sudorifics* (London: Edward and Charles Dilly, 1768), 161 ff.

18. Forster, *Treatise*, 254–55. Forster also noted the healing powers of cold air and wine: "[I]ndeed those who 'indulge in wine,' to such a degree as keeps up the due quantity of perspiration, may escape colds or fevers: whether at the expense of frugality and decency is not my affair to discuss," 355.

19. Patrizia Calefato, *The Clothed Body* (Berg: Oxford, 2004). See also Martin Evans, Dani Cavallaro, and Alexandra Warwick, *Fashioning the Frame: Boundaries, Dress, and the Body* (Berg: Oxford, 1998).

20. Forster, *Treatise*, 355.

21. Anne Buck, *Dress in Eighteenth-Century England* (London: B.T. Batsford, 1979), 187.

22. A.P. Wadsworth and J. de L. Mann, *The Cotton Trade and Industrial Manchester* (Manchester, [s.n.] 1931), 133.

23. Margaret Spuford, "Fabric for Seventeenth-Century Children and Adolescents' Clothes," *Textile History* 34 (2003): 47–63.

24. Gilbert White, *The Natural History of Selborne* (London: T. Bensley, 1789), 189.

25. Buck, *Dress*, 194–200. Phyllis Deane has found that "during the last three decades of the century, although exports of woolen manufactures expanded faster than ever, the industry's total consumption of raw material expanded at a rate that was well below the rate of population growth. The implication is that domestic consumption per head of the population actually declined," in Phyllis Deane, "The Output of British Woollen Industry in the Eighteenth Century," *Journal Of Economic History* 17 (1957): 207–223; 222; see also George Daniel Ramsay, *The English Woolen Industry 1500–1750* (Basingstoke: MacMillan, 1982); Carole Shammas, "The Decline of Textile Prices in England and British America Prior to Industrialization," *The Economic History Review* 47 (1994): 483–507; and John Smail, *Merchants, Markets, and Manufacture: The English Wool Textile Industry in the Eighteenth Century* (Basingstoke: Macmillan, 1999).

26. [John Patteson for] the Select Committee on State of Woollen Manufacture of England, *Report and Minutes of Evidence on the State of Woollen Manufacture of England* (London: House of Commons

Papers 268 and 268a, III 595, 1806). On the slump in woolen trade, see Anne Puetz, "Design Instruction for Artisans in Eighteenth-Century Britain," *Journal of Design History* 12 (1999): 217–239; Beverley Lemire, "Developing Consumerism and the Ready-Made Clothing Trade in Britain, 1750–1800," *Textile History* 15 (1984): 21–44.

27. William Thomson, *Prospects and Observations on a Tour in England and Scotland* (London: G.G.J. and J. Robinson, 1791), 189.

28. John Luccock, *The Nature and Properties of Wool, Illustrated; with a Description of the English Fleece* (Leeds: E. Baines, 1805); idem, *An Essay on Wool, Containing a Particular Account of the English Fleece* (London: J. Harding, 1809); Robert Bakewell, *Observations on the Influence of Soil and Climate upon Wool* (London: J. Harding, 1808); on the need for self-sufficiency of national wool industry, see Society for the Improvement of British Wool (Edinburgh), *Observations on the Advantages Which the Public May Expect to Derive, by Means of the Proposed Association for the Improvement of British Wool* (London: [n.p.], [1790]). Highland Society of Scotland, *Report of the Committee to Whom the Subject of Shetland Wool Was Referred* (Edinburgh: W. Creech, 1790). Other discussions on the issue include John Anstie, *A Letter Addressed to Edward Phelips, Esq. Containing General Observations on the Advantages of Manufacturing the Combing Wool of England* (London: Stafford and Davenport, 1788), idem, *A Letter to the Secretary of the Bath Agriculture Society on the Subject of a Premium, for the Improvement of British Wool* (London: George Stafford, 1791); Henry Wansey, *Wool Encouraged without Exportation* (London: T. Cadell, 1791). Concerns over the imports of Irish wool and linens were the subject of *The Petition of the Merchants, Manufacturers, and Others Concerned in the Wool and Woollen Trade, of Great Britain* (London: W. Phillips, [1800]); for a riposte, see John Holroyd Sheffield, *Observations on the Objections Made to the Export of Wool from Great Britain to Ireland* (London: Debrett, 1800).

29. John Sinclair, *The Statistical Cccount of Scotland* (vol. 7, Edinburgh: W. Creech 1791–99), 161.

30. Ferriar, *Medical Histories and Reflections*, vol. 3; Ferriar advocated the establishment of "clothes-clubs" for secondhand garments. See also Steven King, "Reclothing the English Poor, 1750–1840," *Textile History* 33 (2002): 37–47; John Styles, "Clothing the North: The Supply of Non-Elite Clothing in the Eighteenth-Century North of England," *Textile History* 25 (1994): 139–166; on clothing clubs, see Peter Jones, "Clothing the Poor in Early Nineteenth-Century England," *Textile History* 37 (2006): 17–37.

31. William Cobbett, *Political Register*, December 23, 1815.

32. Thornton, *The Philosophy of Medicine*, vol. 1, 160; Renbourn, "The History of Flannel Binder and the Cholera Belt."

33. Ball, *A Treatise of Fevers*, 42. Edward Strother, in *The Family Companion for Health or, the Housekeeper's Physician* (London: John and James Rivington, 1750), noticed the effects that the practice had on consumptives.

34. Wainewright, *A Mechanical Account of the Non-Naturals*, 172–73. On the inconvenience of flannel and overheating, see *The Best and Easiest Method of Preserving Uninterrupted Health to Extreme Old Age* (London: R. Baldwin, 1748), 152. More criticism appears in *The Best and Easiest Method of Preserving Uninterrupted Health to Extreme Old Age* (London: R. Baldwin, 1748), 159 ff, where the authors sets to discredit the enfeebling effects of overdressing in wool.

35. Wainewright, *Mechanical Account*, 142.

36. Sutherland, *Attempts to Revive Antient Medical Doctrines*, vol. 2, 137–8; Browne Langrish, *The Modern Theory and Practice of Physic* (London: L. Hawes, 1761), 265. Langrish wrote that "it relaxes the subcutaneous Glands, and their excretory ducts, will consequently promote perspiration and sweat; and since it is generally advised to weakly people, who vessels are already too lax, and who sweat too much, it must be seen as a very pernicious custom. The common excuse for it is, that it dries up the sweat, and prevents their catching cold; but this is a great deceit, and so inconsistent with the laws of animal economy, that any one who is in the least versed in them may easily discover the fallacy. A cold-bath will more effectually prevent catching cold, than a flannel shirt." See also David Tyrell and Michael Fielder, *Cold Wars: The Fight Against the Common Cold* (Oxford: Oxford University Press, 2002).

37. Buchan, *Domestic Medicine*, 112. On hardening, see Thomas Reid, *Directions for Warm and Cold Sea-Bathing* (Dublin: William Gilbert, 1795).

38. *The Best and Easiest Method of Preserving Uninterrupted Health*, 159 ff.

39. Jonas Hanway, *Advice from a Farmer to His Daughter, in a Series of Discourses, Calculated to Promote the Welfare and True Interest of Servants* (London, 1770), 349 ff.

40. John Armstrong, *The Art of Preserving Health* (Dublin: James Rudd, 1756), 104–5.

41. Forster, *Treatise*, 121–22. Flannel waistcoats were in high esteem even in the warmest parts of Europe, argued Jonas Hanway: "When people wear their fines dress, they take care to be the warmest; whereas the contrary is practiced among our gentry [...] I suppose it is for this reason that so many of the gentry die at an early age. Ten of them are carried off by consumption, to one of us. Perhaps we enjoy some advantages, by living more under the canopy of the heavens; though God knows, often exposed in the extreme, to hot and cold, dry and moist weather," Jonas Hanway, *Virtue in Humble*

Life: Containing Reflections on the Rreciprocal Duties of the Wealthy and Indigent (vol. 2, London: Dodsley, 1777), 184.

42. On swaddling, see Nicolas Andry de Bois-Regard, *Orthopædia: or, the Art of Correcting and Preventing Deformities in Children* (London: A. Millar, 1743). On "bracing" environments, see also John Scot, *An Enquiry into the Origin of the Gout.* (London: the Author, 1783), 188 ff; Thornton, *The Philosophy of Medicine*, vol. 2, 40; see also Morwena and John Rendel-Short, *The Father of Child Care: Life of William Cadogan (1711–1797)* (Bristol: John Wright, 1966); Reinhard Spree, "Shaping the Child's Personality: Medical Advice on Child-Rearing from the Late Eighteenth to the Early Twentieth Century in Germany," *Social History of Medicine* 5 (1992): 317–35.

43. Cold and wet feet of the well-bred children, he says, are no worse than the cold and wet feet of the poor. For the poor, there's no difference between cold feet and cold hands: "[I]f someone did protect their hands in the same way people protect their feet, wet hands would become equally prejudicial to health. And what is it that makes this great Difference between the Hands and Feet in others, but only Custom?" Locke opposed waterproofing: His ideal shoes leaked, and cold water ought to be applied to children's feet every night, gradually, starting with lukewarm water in spring, and then becoming colder during the summer months. For iconography of feet in early modern Europe, see Margaret Pelling, "The Body Extremities: Feet, Gender, and the Iconography of Healing in Seventeenth-Century Sources," in Hilary Marland and Margaret Pelling (eds.), *The Task of Healing: Medicine, Religion, and Gender in England and the Netherlands, 1450–1800* (Rotterdam: Erasmus Publishing, 1996), 221–251.

44. Although associated with a stoic approach to child care, Underwood favored moderation: "I have seen a child of four years old, the daughter of people of fashion, whose legs were covered with these sores up to the knee, and yet the lady could not be prevailed upon in time, to suffer stockings to be put on, because strong and healthy children are thought to be better without them," Michael Underwood, *A Treatise on the Disorders of Childhood, and Management of Infants from the Birth* (vol. 3, London: J. Matthews, 1797), 50.

45. William Cadogan, *An Essay upon Nursing and the Management of Children* (London: Robert Horsfield, 1769), 9–10. Also see John Edmonds Stock, *An Inaugural Essay on the Effects of Cold upon the Human Body* (Philadelphia: Joseph Gales, 1797).

46. Cadogan *Essay*, 10.

47. Jules Barbey D'Aurevilly (tr. George Walden), *Who's a Dandy: Dandyism and Beau Brummell* (London, 2002).

48. C. Willet and Phillis Cunnington, *The History of Underclothes* (London: Michael Joseph, 1951), 97 ff. Daniel Roche, *The Culture of Clothing: Dress and Fashion in the Ancien Regime* (Cambridge: Cambridge University Press, 1994), 504.

49. University of Edinburgh encouraged dissertations on the topic: Joannes Hannum Gibbons, *Medica Inauguralis, Quaedam de Vestitu Laneo* (Edinburgi: Balfour et Smellie, 1786); Samuel Hughes, *De Vestitu* (Edinburgi: Apud Murray et Cochrane, 1795).

50. Margaret Hunt, "Racism, Imperialism, and the Traveler's Gaze in Eighteenth-Century England," *The Journal of British Studies* 32 (1993): 333–57, 344.

51. Casey Finch, "Hooked and Unbuttoned Together: Victorian Underwear and Representations of the Female Body," *Victorian Studies* 34 (1991): 337–63, 339.

52. Thornton, *Philosophy of Medicine*, 47.

53. Alexander. *Experimental*, 183; Adair, *Essays on Fashionable Diseases*, 67; on hiccups, see William Rowley, *The Rational Practice of Physic* (4 vols., London: E. Newberry, 1793), vol. 2, 290; Lewis Mansey, *The Practical Physician; or, Medical Instructor* (London: W. Stratford, [1800]), 140.

54. Gerard Freiherr van Swieten, *The Commentaries upon the Aphorisms of Dr. Herman Boerhaave* (London: Robert Horsfield, 1744–1773), vol. 17, 519.

55. Millar, *Observations on the Prevailing Diseases in Great Britain*, 301.

56. Adair, *Commentaries*, 229.

57. For the uses of flannel in tropical climates, see William Hillary, *Observations on the Changes of the Air and the Concomitant Epidemical Diseases, in the Island of Barbadoes* (London, L. Hawes, 1766). Further support of this view is in Benjamin Moseley, *A Treatise on Tropical Diseases; on Military Operations; and on the Climate of the West-Indies* (3rd ed., London: T. Cadell 1792), 190–193. Hector McLean, *An Enquiry into the Nature, and Causes of the Great Mortality among the Troops at St. Domingo* (London: Cadell, 1797). McLean was cautious to refer to adverse effects of its prolonged use in hot atmospheres, such as a general debility of system, sapping of the energy of the soldier, skin afflictions, and stench, 269.

58. *Times*, November 30, 1785, 4; Millinery Rooms in Covent Garden advertised for the new season, among other items, calico flannel, so "salutary and pleasant for under dresses." *Times*, December 11, 1787, 1. The literature on domestic health dealing with flannel included Cornwell, *The Domestic Physician*, 431, discussing flannel in relation to diarrhea and rheumatism; Andrew Harper, *The Oeconomy of Health, or, a Medical Essay: Containing New and Familiar Instructions for the Attainment of Health, Happiness, and Longevity* [London: C. Stalker], [1785?], 26; John Grigg, *Advice to the Female Sex in General* (Bath: S. Hazard, 1789), 64, recommends it for menstrual disorders; Nicholas Culpeper, *Culpeper's English Family Physician* (London: W. Locke, 1792), vol. 2, 279.

59. Thomas Hayes, *A Serious Address on the Dangerous Consequences of Neglecting Common Coughs and Colds* (2nd ed., London: J. Murray),

1786; five editions were published from 1783 to 1808. Hayes criticized William Heberden's view on the benignity of damp linen: "I am sorry that so a deservedly great physician as Dr Hebberden [sic] should even doubt that wet rooms, damp beds, wet linen etc should not occasion illness to human body, when it has not only be recorded by physicians from the earliest ages to the present day," 67 n.

60. Hayes, *A Serious Address*, 56–7.
61. On thin clothes, see Elizabeth Ewing, *Dress and Undress: A History of Women's Underwear* (London: Batsford, 1978).
62. Hayes, *A Serious Address*, 55.
63. Hayes, *A Serious Address*, 57, 62. Gilbert Blane wrote, "[M]ay not clothing be considered as a *filter* as it were to separate the impurities of the air before it comes in contact with the surface of the body and, therefore, sudden and unreasonable change of apparel be very unsafe for health."
64. Sir Benjamin Thompson, "Experiments Made to Determine the Positive and Relative Quantities of Moisture Absorbed from the Atmosphere by Various Substances, under Similar Circumstances, Communicated by Charles Blagden," *Philosophical Transactions of the Royal Society* 77 (1787): 240–245; for Thompson's interest in clothing see Ann M. Little, "Shoot that Rogue for He Hath an Englishman's Coat On: Cultural Cross-Dressing on the New England Frontier," *The New England Quarterly* 74 (2001): 239–73, 242–43.
65. Thompson, "Experiments," 243.
66. Thompson, "Experiments," 244.
67. Thompson, "Experiments," 244.
68. Adair, *Essays on Fashionable Diseases*, 147.
69. William Buchan, *A Letter to the Patentee, Concerning the Medical Properties of Fleecy Hosiery*. (3rd ed., London: Peterborough-House Press, 1790).
70. Buchan, *Letter*, 13. Buchan insisted on the medical meaning of the product: "On a review of those diseases in which flannel is known to act, either as a preventive or remedy, they will be found more numerous than all that are cured by any one article of *Materia Medica*." Ibid, 14.
71. Walter Vaughan *An Essay, Philosophical and Medical, Concerning Modern Clothing*. [Rochester], 1792, 6.
72. Vaughn, *Essay*, 1.
73. Vaughan, *Essay*, 23.
74. "I do not mean to say that fashions have not originated from the blemishes of individuals in power. I know they have. But I deny that any man's having a blemish and using means to hide it, is a sufficient cause for my using the same means who have not that blemish." Vaughan, *Essay*, 29.
75. Vaughan, *Essay*, 38.

76. Vaughan, *Essay*, 103.

77. Vaughan, *Essay*, 98. Sailors, soldiers, and husbandmen lacked these times and opportunities, and "as the necessity of these people must always have existed it does not seem likely that such a covering should have been intended for them by our all wise Creator as they have neither ability nor opportunity to wear," 96. For critique of frequent changing of apparel, see Thornton, *Medical Extracts,* vol. 2, 267. Referring to Beddoes, Thornton asked: "[I]s it not alone their delicacy of constitution, or their being more confined within doors, but the frequent changes they make in the quality of their garments, and sometimes, however fearful of a partial current of air, because they exposed even those parts of the body, that a little before had been warmly covered," 268.

78. Willich, *Lectures on Diet and Regimen*, 265–6. Among the more elaborate apologies of providential argument was Thornton, *Philosophy of Medicine*, 41–42.

79. Willich, *Lectures*, 270.

80. Willich, *Lectures*, 267–84. Johann Wilhelm von Archenholz wrote in his *Picture of England* (Dublin: P. Byrne, 1790), 330: "The women of the middle rank, when it is dirty weather, fix on their feet certain circles of iron, which raise them above the dirt, and are tied to their shoes with latches, so that they are not wetted. The custom of wearing shoes only of silk or cloth, (for a common servant will not put on a leather shoe), has undoubtedly given occasion to theses strange machines which they put off at the door, when they enter the house, and which besides their unavoidable inconvenience, have besides the disadvantage of producing an awkward and disagreeable gait."

81. James Parkinson, *The Villager's Friend and Physician: or the Familiar Address on the Preservation of Health* (London: James Sammels, 1800), 30.

82. *The Mechanics' Chronicle* (London: London Mechanics Institution, 1824), 16.

83. Archibald Cochrane, Earl of Dundonald, *A New Year's Gift, Recommended to the Perusal of Persons Desirous of Promoting Objects of Domestic Economy, Their Own Health* (London: n.p. 1798), 10. Also noted was the decline in neoclassical cut, in France at least: "A considerable alteration for the better has within a short time taken place in the dress of the Parisian Ladies, and their health, as well as their manners, will be much improved by the substitutions which they continue to make for the fashions of antiquity. It is remarked with pleasure, that the gowns and robes are in general less open about the bosom, while the women of character have abandoned [... a mode of dress that shocked every feeling of decorum and property," *Times*, April 27, 1802; 2.

84. William Stevenson, *General View of the Agriculture of the County of Surrey* (London: R. Philips, 1809), 364; Andrew Swinton, *Travels*

into Norway, Denmark, and Russia, in the Years 1788, 1789, 1790, and 1791 (London, J. Robinson, 1792), 434.

85. On Nott, see *Times*, November 3, 1791. Flannel waistcoats were recommended by Sir John Sinclair, *Essays on Miscellaneous Subjects* (London: A. Strahan,1802), 423. For use among the sugarcane plantation staff and slaves against "catching of cold" after working near boilers, see Dr. Collins, *Practical Rules for the Management and Medical Treatment of Negro Slaves, in the Sugar Colonies* (London: J. Barfield 1803), 130. For proposed convict apparel in winter months, see Robert Edington, *A Descriptive Plan for Erecting a Penitentiary House, for the Employment of Convicts* (London: A. Topping, [1803?]), 29. For friction's benefits, see Willich, *The Domestic Encyclopaedia*, 25. For Count Berchtold's advice for travelers using flannel waistcoats in all weather, see *The Cheap Magazine* (Haddington: George Miller and Son, 1814), 284. For averting common cold on journeys, see *The Traveller's Oracle: or, Maxims for Locomotion* (London: H. Colburn, 1827), 13. For the use in laborers' daily attire, see Country Gentleman, *Observations on the Corn Trade, Agriculture, and Manufactures, of England* (London: J. Ridgway, 1815), 7. On how flannel saves money that the poor otherwise spent on medicines, see John Conolly, *The Workingman's Companion: Cottage Evenings* (London: C. Knight, 1831), 20; [Anon.], *The Advantage of Warm Clothing*: (London: Jarrold, [1857], Copy 1 Wellcome Library Supplier/Donor: T. & M. From Jarrold's Household Tract Series).

86. On chest cramps attributed to having one's waistcoat taken off for washing, see Henry Hunt, *To the Radical Reformers of England, Ireland, and Scotland* (London: W. Molyneux, 1820–22), vol. 1, 20. On a journeymen carpenter catching fatal cold, see Joshua Kirby Trimmer, *Further Observations on the Present State of Agriculture, and Condition of the Lower Classes of the People, in the Southern Parts of Ireland:* (London: F.C. and J. Rivington, 1812), 80.

87. "Subscription for Providing the British Troops on the Continent with Worsted Stockings and Flannel Caps. Address to the Public," See *Times*, November 8, 1793, 2.

88. "Advertisement: Duke of York's Army. Flannel Waistcoats, Drawers, Stockings, Caps, and Blankets," *Times*, November 28, 1793, 2. One 1798 list of clothing articles for regiments of cavalry and infantry included "coat, waistcoat, plush breeches, flannel waistcoat, flannel drawers, watering cap, hat, gloves," see House of Commons, Select Committee on Finance. *Twenty-Third (Thirty-Sixth) Report from the Select Committee on Finance* (n.p., 1798), vol. 2, 114; see also Thomas Reid, *A Treatise on Military Finance, Containing the Pay and Allowances in Camp, Garrison, and Quarters of the British army.* (9th ed., London: T. Egerton, 1805), vol. 2, 21, 33.

89. *Liberty Scraps* (London: n.p. 1794), 16.

90. Beddoes, *Manual*, 47.

91. Beddoes, *Manual*, 222.

92. Beddoes, *Manual*, 224.

93. A review of cold-induced ills in contemporary physiological jargon is exemplified by Henry Clutterbuck's "Lectures on the Theory and Practice of Physic," *Lancet* 5 (1826): 785–789. Clutterbuck was one of the founders of the London Meteorological Society in 1823. See also William Frederick Barlow, "On the Effect of Temperature in Causing Reflex Action," *Lancet* 40 (1843): 219–221; Henry Searle, "On the Agency of the Atmosphere on the Production of Disease," *Lancet* 24 (1835): 110–114. In therapy, cold had seen many applications in wound treatment, midwifery, anesthesia, insanity, and cold-water bathing.

94. Articles on the role of flannel in health during the first half of the nineteenth century include J.D., "On Preserving the Health of Manufacturers," *Tradesman, or, Commercial Magazine* 5 (1810): 218–224; Medical Practitioner, "Medical Report," *Newcastle Magazine* 3:12 (1824) 610–616; Andrew Combe, *The Principles of Physiology Applied to the Preservation of Health* (London: Longman, 1834); "Chemistry of Winter," *Chambers's Edinburgh Journal* 259 (1848): 387–341.

95. See "George Lefevre Biography," *Literary Gazette* 1518 (1845): 180.

96. Sir George Lefevre, *Thermal Comfort: or Popular Hints for Preservation from Colds, Coughs, and Consumption* (Second edition, London: John Churhill, 1844), 9. On Lefevre, see Elizabeth Baigent, "Sir George William Lefevre (1798–1846)," *Oxford Dictionary of National Biography*.

97. S.M. Newton, *Health, Art, and Reason: Dress Reforms of the Nineteenth Century* (London: J. Murray, 1974); P.A. Cunningham, *Reforming Women's Fashion, 1850–1920* (Kent, OH: Kent State University Press, 2003).

98. Beddoes, *Manual of Health*, 220.

99. Adair, *Essays on Fashionable Diseases*, 66.

100. For a different but pertinent approach, see Christopher Lawrence and Steven Shapin (eds.), *Science Incarnate: Historical Embodiments of Natural Knowledge* (Chicago: University of Chicago Press, 1998).

101. Peter Conrad, "Medicalization and Social Control," *Annual Review of Sociology* 18 (1992): 209–32.

5   The Choice of Air

1. Edward Morris, *The Secret, A Comedy* (Dublin: G. Walsh, 1799), 18.

2. Hanway, *Virtue in Humble Life*, vol. 2, 272.

3. A legal meaning of the term appeared in Michael Dalton's early modern discussion on the curbing powers of vagrancy laws, which maintained that "[a] man without his offence is barred of his natural liberty, [and] at the same time deprived of the company of friends and relations, choice of air, and place of trade." See Michael Nolan, *A Treatise of the Laws for the Relief and Settlement of the Poor*, vol. 1 (London: A. Strahan, 1805), 146; the original work is Michael Dalton, *The Country Justice: Containing the Practice, Duty and Power of the Justices of the Peace* (London: E. and R. Nutt, 1727), 231 and the reference to 13 & 14 Car. II c.12. The "change of air" also had a technical meaning: "These Soldiers having but very little Fatigue, the Officers found Means to make them perform all the Duty that was required of them, while they remained in Garrison; but upon their being brought home, the Change of the Air, added to their own Age and Infirmities, made most of them entirely unfit for a March, or for any Duty whatever." William Pulteney to H. Pelham, "First Parliament of George II: Fifth Session (part 2 of 4, from 28/1/1732)," *The History and Proceedings of the House of Commons* 7: (1742): 133–59.

4. Arbuthnot, *Essay*, 205.

5. *The Best and Easiest Method of Preserving Uninterrupted Health to Extreme Old Age* (London: R. Baldwin, 1748).

6. Maximilian Hazlemore. *Domestic Economy; or, a Complete system of English Housekeeping* (London: J. Creswick, 1794), 357.

7. Thomas Salmon, *The History and Present State of the British Islands* (2 volumes, London: J. Robinson, 1743), vol. 2, 203.

8. Bath, *An Essay on the Medical Character,* 134.

9. Ebenezer Gilchrist, *The Use of Sea Voyages in Medicine* (London: A. Millar, 1757), xi.

10. "Proper choice of air and aliments" is mentioned in Antoine-François de Fourcroy, *Elements of Natural History and Chemistry* (London: C. Elliot and T. Kay; and Edinburght: C. Elliot, 1790), 13; see also John Hill, *The Old Man's Guide to Health and Longer Life* (Dublin: James Hoey, 1760), 20. Soranus, Coelius, and Methodists were identified as attentive in their "choice if the air for the sick, as in directing a large cool chamber for them, in hot inflammatory diseases on a hot season," William Hillary, *An Inquiry into the Means of Improving Medical Knowledge* (London: C. Hitch and L. Hawes, 1761), 105; Friedrich Hoffmann, *A Treatise of the Extraordinary Virtues and Effects of Asses Milk* (London: John Whiston and Benjamin White, 1754), 2; "the choice of Air is undoubtedly of the utmost consequence to moderate and reduce a constitution when it grows into any intemperature," Andrew Hooke, *An Essay on Physick. Or, an Attempt to Revive the Practice of the Antients* (London: J. Roberts, 1734), 137; Langrish, *The Modern Theory and Practice of Physik*, 317; *The Modern Family Physician* (London: F. Newberry,

1775), 95; "the choice of the air is extremely necessary, because it extends its influence to the brain as well as to the body, and enlivens the mind, while it promotes the health," Thomas Moffett, *Health's Improvement: or, Rules Comprizing and Discovering the Nature, Method and Manner of Preparing all Sorts of Foods Used in This Nation* (London: T. Osborne, 1746), 83; see also Nicholas Robinson, *An Essay on the Gout* (London: Edward Robinson, 1755), 139. The architectural view on the problem is exemplified in Isaac Ware, *A Complete Body of Architecture* (London: J. Rivington, 1768), 97. On the popular view of "air-change," see Wayland D. Hand, "Folk Medical Inhalants in Respiratory Disorders," *Medical History* 12 (1968): 153–167.

11. The health bard John Armstrong immemorialized "the choice of aliment, the choice of air" in *The Art of Preserving Health*, 69. The translation of Jesuit Leonard Lessius's *Hygiasticon* rendered the author's confession as of someone who "was ever very nice also in the choice of Air, and took all imaginable Care to screen myself from the Inclemency of the Winds and Weather," Leonardus Lessius, *Hygiasticon: or, a Treatise of the Means of Health and Long Life* (London: the author, 1742), 14. Other practitioners proposed fine-grained taxonomies of air that corresponded to constitutional idiosyncrasies: "[I]t should be observed that generally speaking, consumptive Patients, such as before described, that are emaciated, and of a hot Temperament, make an injudicious Choice in changing their Air for that which is more refined, especially if they are given to spit blood; such patients should avoid the hot and subtle air of the South of France, but should fly for refuge to the United Provinces, where they will find the Air of the Hague and Utrecht more beneficial to them than that of Clermont and Montpelier; and on the other Hand, gross and corpulent persons encumbered with a load of Humours, who are likewise affected [...] may reasonably expect to find their Account by removing to the Hills in their own Country." Richard Blackmore, *A treatise of Consumptions and Other Distempers* (London: John Pemberton, 1724), 127–8.

12. Thomas Curteis, *Essays on the Preservation & Recovery of Health* (London: Richard Wilkin, 1704), 10; Buchan, *Domestic Medicine*, 417.

13. Sutherland, *Attempts*, vol. 2, 50.

14. Gregory, *A Dissertation on the Influence of Change of Climate in Curing Diseases*, 18. On disease causation, see Hamlin, "Predisposing Causes," 1992.

15. Falconer, *Remarks on the Influence of Climate*, 5; Ann Borsay, "William Falconer," in *Oxford Dictionary of National Biography* (Oxford: Oxford University Press, 2004).

16. Gregory, *Dissertation*, 36. Such an approach made warmth a logical choice: "[T]he excellency of southern air consists in its warmth

and constancy. The pores are constantly open; thus are the lungs constantly relieved," Sutherland, vol. 2, 46. Sutherland also wrote: "[S]ometimes we learn that consumptives grow worse in pure, serene, dry air; and that they mend in seasons seemingly unfavourable. Moist air bids fairer for preserving lungs from inflammation than dry. It refreshes and ventilates the blood. How comfortable is the shower in great heats!" Judging local topography, in his view, involved vulgar error illustrated the custom of strangers to avoid accommodation near the river Avon. They made their judgment on purely olfactory grounds, smelling its vapors during low tide, but "the banks of the river Avon, are constantly refreshed by the flux and reflux of the tides. The mud is saturated with particles saline and salutary. That air which enters by the back windows of the Well-House is drying, penetrating, and, nature's balsam," 43.

17. Gregory, *Dissertation*, 32. Sutherland wrote: "People inhabiting great cities or low confined situations, are obliged to change the air for the preservation of life, or the cure of disease [...] Instead of the artificial applications of antiseptics, balsamics, heaters, and dryers, the ancients directed consumptives to medicated air." Sutherland, *Attempts*, 50.

18. Sutherland, *Attempts*, 132.

19. William Heberden, *Commentaries on the History and Cure of Disease* (Boston: Wells and Lill, 1818), 307.

20. Edward Strother, *An Essay on Sickness and Health* (London: Charles Rivington, 1725), 8; which was used in Forster's *A Treatise on the Causes of Most*, 250.

21. Review of Edward Kentish, *Account of Baths, and of a Madeira House at Bristol* (Longman: London, 1815), in *The Medical and Physical Journal* (1815): 61–65. The establishment was favorably mentioned in Robert Buchanan, *A Treatise on the Economy of Fuel and Management of Heat* (London: the author, 1815), 304, and John Evans, *The Picture of Bristol* (Bristol: W. Sheppard, 1818), 85; see also Eneas MacKenzie, *A Descriptive and Historical Account of the Town and County of Newcastle upon Tyne* (Newcastle upon Tyne: Mackenzie and Dent, 1827), 506n.

22. For tropical "imaginary geography," see J. S. Duncan, "The Struggle to Be Temperate: Climate and 'Moral Masculinity' in Mid-Nineteenth Century Ceylon," *Singapore Journal of Tropical Geography* 21 (2000): 34–47; for English landscapes, see Y. Shu-Chuan, *"Spinning about the World": Imagined Geographies in Elizabeth Gaskell's major fiction.* PhD thesis, University of Manchester, 2003. On imaginary space, see Edward Soja, *Thirdspace: Journeys to Los Angeles and Other Real-and-Imagined Places* (Oxford: Blackwell, 1996). On travel and cultures of tourism, see O. Hamilton, *Paradise of Exiles: Tuscany and the British* (London: Andre Deutsch, 1974), H. Berghoff, B. Korte, B. R. Schneider, C. Harvie (eds.), *The Making of Modern*

Tourism: The Cultural History of the British Experience, 1600–2000 (London: Palgrave Macmillan, 2002).

23. Discussions on climate and health in the Montpellier School of Medicine include Jean Louis Leno Marchant, *Dissertation sur les Voyages consideree comme moyen theraputique* (Montpellier: Jean Martel, 1816); E.D. Sauveur Vader, *L'influence des climats sur l'homme* (Montpellier: Jean Martel, 1814).

24. On varieties of early modern European travel, see Jozsef Borocz, "Travel Capitalism: The Structure of Europe and the Advent of the Tourist," *Comparative Studies in Society and History* 34 (1992): 708–741. For comprehensive treatments of eighteenth-century travel, see Percy G. Adams, *Travelers and Travel Liars, 1660–1800* (Berkeley: University of California Press, 1962); Charles L. Batten, Jr., *Pleasurable Instruction: Form and Convention in Eighteenth-Century Travel Literature* (Berkeley: University of California Press, 1978).

25. Peregrine Horden, "Travel Sickness: Medicine and Mobility in the Mediterranean from Antiquity to the Renaissance," in W.V. Harris (ed.), *Rethinking the Mediterranean* (Oxford: Oxford University Press, 2005), 179–99, 188 ff; see also Peregrine Horden and Nicholas Purcell, *The Corrupting Sea: A Study of Mediterranean History* (Oxford: Malden, MA, 2000). On medieval discussion of the salutary effects of Palestine as the ideal site for the reestablishment of the Jewish people, see Stephen T. Newmyer, "Climate as Science and Metaphor in the Writings of Jehuda Halevi," in Janet Perez and Wendell Aycock (eds.), *Climate and Literature: Reflections of Environment* (Detroit: Alvin Snider, 1997), 19–28; more generally, see Irina Metzler, "Perceptions of Hot Climate in Medieval Cosmography and Travel Literature," *Reading Medieval Studies* 23 (1997): 69–105.

26. Karen Ordahl Kupperman, "Fear of Hot Climates in the Anglo-American Colonial Experience," *The William and Mary Quarterly* 41 (1984): 213–40; Alan Bewell, "Jefferson's Thermometer: Colonial Biogeographical Constructions of the Climate of America," in Noah Heringman (ed.), *Romantic Science: the Literary Forms of Natural History* (Albany: New York State University Press, 2003), 118–138; Chris Tiffin, "Imagining Countries, Imagining People: Climate and the Australian Type," *Span* 24 (1987): 46–62. For a review on southern inferiority, see J.W. Johnson, "Of Differing Ages and Climes," *Journal of the History of Ideas* 21 (1960): 465–480, and Felicity A. Nussbaum, *Torrid Zones: Maternity, Sexuality, and Empire in Eighteenth-Century English Narratives* (Baltimore: Johns Hopkins University Press, 1995).

27. See Caroline Hannaway, "The Societe Royale de Medecine and Epidemics in the Ancient Regime," *Bulletin of Medical History* 46 (1972): 257–273. On the general notion of climate, health, national customs, and character, see Glacken, *Traces on the Rhodian Shore,*

chapter 12. On folk inhalants and the popular "change of air," see Hand, "Folk Medical Inhalants," 1968.

28. Alain Corbin, *The Lure of the Sea: The Discovery of the Seaside in the Western World*, 1750–1840 (Cambridge: Polity Press, 1994).

29. R. S. Neale, *Bath 1680–1850: A Valley of Pleasure Yet a Sink of Iniquity* (London: Routledge & Kegan Paul, 1981), 1–26; J. Walvin, *Beside the Seaside: A Social History of the Popular Seaside Holiday* (London: Allen Lane, 1978), 11–33; Lencek, and G. Bosker, *The Beach: The History of Paradise on Earth* (Harmondsworh: Penguin, 1999).

30. See Jeremy Black, *The British Abroad: The Grand Tour in the Eighteenth Century* (London: Sutton Publishing, 1992); J. Buzard, *The Beaten Track: European Tourism, Literature, and the Ways to "Culture,"* 1800–1918 (Oxford: Clarendon Press, 1993); John Spillane, *Medical Travellers: Narratives from the Seventeenth, Eighteenth, and Nineteenth Centuries* (Oxford: Oxford University Press, 1984); Edward Chaney, *The Evolution of the Grand Tour: Anglo-Italian Cultural Relations Since the Renaissance* (Portland, OR: Frank Cass, 1998); For recent views on the "cultured" and social aspects of the Tour that include voyage literature, antiquarianism, sightseeing, and gender roles, see the articles in Grand Tour Forum, *Eighteenth-Century Studies* 31 (1997): 87–112.

31. "Review of the Books on Mineral Waters and Spas," *British Medical and Foreign Review* October 1842, 310.

32. J. A. Mason, *A Treatise on the Climate and Meteorology of Madeira* (London: Churchill, 1850), 111.

33. See E.S. Turner, *Taking the Cure* (London: Michael Joseph, 1967); John Pemble, *The Mediterranean Passion: Victorians and Edwardians in the South* (Oxford: Clarendon Press, 1987); I. Littlewood, *Sultry Climates: Travel and Sex Since the Grand Tour* (London: John Murray, 2000); R. Aldrich, *The Seduction of the Mediterranean: Writing, Art, and Homosexual Phantasy* (London: Routledge, 1993). Charles Sackville of Stoneland wrote in 1788: "I owe all this health that I am boasting of, almost entirely to change of air and change of scene, for I own that Spa is, this summer, one of the most stupid places I ever saw," quoted in Black, *The British Abroad*, 183.

34. Pemble, *The Mediterranean Passion*, 245.

35. William Marcet, *The Principal Southern and Swiss Health Resorts* (New York: Scribner and Welford, 1883), 313.

36. David N. Livingstone, "Tropical Climate and Moral Hygiene: The Anatomy of a Victorian Debate." *British Journal for the History of Science* 32 (1999): 93–110; David Arnold (ed.), *Warm Climates and Western Medicine: The Emergence of Tropical Medicine, 1500–1900* (Rodopi: Amsterdam, 1996); for an overview, see Conevery Bolton-Valencius, "Histories of Medical Geography," in Nicolaas Rupke (ed.), *Medical Geography in Historical Perspective* (London: Wellcome Trust for the History of Medicine, 2000), 3–30.

37. For analogous agnosticism as it related to meteorological explanation, see Vladimir Janković, *Reading the Skies: A Cultural History of English Weather, 1650–1820* (Chicago: Chicago University Press, 2000), Chapter 7.

38. James Clark, *Medical Notes on Climate, Disease, Hospitals, and Medical Schools in France, Italy, and Switzerland* (London: T.G. Underwood, 1820); idem, *Observations on the System of Teaching Clinical Medicine in the University of Edinburgh* (London: T. Davidson, 1827); idem, *The Influence of Climate in the Prevention and Cure of Chronic Diseases, More Particularly of the Chest and Digestive Organs* (London: T. G. Underwood, 1829); idem, *A Treatise on Pulmonary Consumption* (London: Sherwood, Ceilbut, and Piper, 1835).

39. James Clark, *The Sanative Influence of Climate with an Account of the Best Places of Resort for Invalids in England, the South of Europe, etc.* (Philadelphia, A. Waldie, 1841). On Clark's career, see Harley Williams, *The Healing Touch* (London: Right Book Club, 1951).

40. James Johnson, *Change of Air, or the Pursuit of Health and Recreation* (London: S. Highely, 1839), 244. On Johnson, see W.A. Greenhill, revised by Mark Harrison, James Johnson, *Oxford Dictionary of National Biography* (Oxford: Oxford University Press, 2004).

41. Johnson, *Change of Air*, 22.

42. Johnson, *Change of Air*, 2. On disease of civilization, see Roy Porter, "Civilisation and Disease: Medical Ideology in the Enlightenment," in Jeremy Black and James Gregory (eds.), *Culture, Politics, and Society in Britain, 1660–1800* (Manchester University Press, 1991).

43. "Review Article 1," *The British and Foreign Medical Review* (October 1842), 312.

44. J. H. Bennet, "The War and the South of France," *Lancet* 96 (1870): 419; see also Henry Maccormac, *On the Nature, Treatment, and Prevention of Pulmonary Consumption* (London: Longman, 1855). On the theory of tuberculosis, see Carter Codell, "The Germ Theory, Beriberi, and the Deficiency Theory of Disease," *Medical History* 21 (1977): 119–136; Margaret Crawford, "Dearth, Diet, and Disease in Ireland: A Case Study of Nutritional Deficiency," *Medical History* 28 (1984): 151–161.

45. Chloe Chard, "From the Sublime to the Ridiculous: The Anxieties of Sightseeing," in H. Berghoff, B. Korte, and R. Schneider, *The Making of Modern Tourism* (London: Palgrave, 2002); see also Chloe Chard, "Lassitude and Revival in the Warm South: Relaxing and Exciting Travel, 1750–1830," in Richard Wrigley and George Revill (eds.), *Pathologies of Travel*, 179–201.

46. Mary Shelley, *Selected Letters of Mary Wollstonecraft Shelley* (Baltimore: John Hopkins University Press, 1995), 381. For the role of evening in the romantic reading of nature, see Christopher A. Miller, *The Invention of Evening: Perception and Time in Romantic Theory* (Cambridge: Cambridge University Press, 2006), 112–145.

47. Johnson, *Change of Air*, 258, my emphasis.
48. Charles Barham, "On Climate, in Some of Its Medical Aspects." *British Medical Journal* 192 (1860): 659–661, 660, my emphasis. Parenthetically, Barham thought that a British possession abroad would enable more people to share temperate climates not only as medical tourists, but also as the workers who would replace the local labor force. The southern landscapes could thus be accessed by the poor, for whom healing abroad would otherwise be a mockery. By this time, the issue of colonial labor force and the coping with extreme tropospheres had already become a burning issue of foreign policy and economics. See Mark Harrison, *Climates and Constitutions: Health, Race, Environment, and British Imperialism in India, 1600–1850* (Oxford: Oxford University Press, 1999); Michael Osborne, *Nature, the Exotic, and the Science of French Colonialism* (Bloomington: Indiana University Press, 1994).
49. Richard Wrigley, "Pathological Topographies and Cultural Itineraries," in Richard Wrigley and George Revill (eds.), *Pathologies of Travel* (Amsterdam: Rodopi, 2000).
50. Pemble, *Mediterranean Passion*, 9.
51. "Review," *Bentley's Miscellany* 32 (1852): 319.
52. Quoted in Wrigley, "Pathological Topographies," 29; see also C. Hoolihan, "Health and Travel in Nineteenth-Century Rome," *Journal of the History of Medicine and Allied Science*, 1989, 462–85. See also Philip Ayres, *Classical Culture and the Idea of Rome in Eighteenth-Century England* (Cambridge: Cambridge University Press, 1997).
53. Hester L. Piozzi, *Observations and Reflections Made in the Course of a Journey through France, Italy, and Germany* (London: A. Strahan, 1798), 101, 103.
54. T. Bowers, "Reconstituting the National Body in Smolletts' Travels through France and Italy," *Eighteenth-Century Life* 21 (1997): 1–25, 22.
55. Clark, *The Sanative Influence*, 34. For advice on travel dress, see John Ingamels, *A Dictionary of British and Irish Travelers in Italy* (New Haven and London: Paul Mellon Centre, Yale University Press, 1997), L (i.e. 50).
56. Tobias Smollett, *Travels through France and Italy* (London: R. Baldwin, 1766), 205–7. The negative opinions took into account a grander view of inconveniences. Rear Admiral William Henry Smyth, vice president of the Royal Society and president of the Royal Geographical Society, warned that "when the sultry and withering blaze of heat, the earthquakes, hurricanes, diseases, misery, personal insecurity, reptiles, mosquitoes, flies, fleas, and other major and minor evils are recollected, the pleasure of visiting warm climates is considerably alloyed," W.H. Smyth, *The Mediterranean. A Memoir Physical, Historical, and Nautical* (London: J.W. Parker, 1854), 263.

On hazards of travel, see Pemble, *The Mediterranean Passion*, 18–38; on unsanitary resorts, see W.S. Playfair, "The Insanitary Condition of Continental Health Resorts," *Lancet* 123 (1884): 273.

57. Patrick Brydone, *A Tour through Sicily and Malta*. (2 volumes, Dublin: J. Potts, 1774), 3.

58. See Pemble, *Mediterranean Passion*, 240–5.

59. Anonymous, "Art III." Sketches of Brazil, etc. *Edinburgh Medical and Surgical Journal* 1865, 218–238, 226.

60. Beddoes, *Manual of Health*, 328; Rosalie Stott, "Health and Virtue: Or How to Keep out of Harm's Way. Lectures on Pathology and Therapeutics by William Cullen, c. 1770," *Medical History* 31 (1987): 123–142.

61. Beddoes, *Manual of Health*, 329–336.

62. Smyth, *The Mediterranean*, 253.

63. Alexander Taylor, *Climates for Invalids, or a Comparative Enquiry as to the Prevention and Curative Influences of the Climate of Pau*. (London: Churchill, 1866), 139.

64. S. Brown, *An Inaugural Dissertation on the Bilious Malignant Fever* (Boston: Manning and Loring, 1797), 52, 53.

65. Henry Matthews, *The Diary of an Invalid being a Journal of a Tour in Pursuit of Health* (Fifth edition, Paris: A. and W. Galignani and Co., 1836 (1821)), 22.

66. Hennen, *Medical Topography*, xxxv, 606.

67. Phthisis (pseud.), "On the Comparative Advantages of Southern Climates in Cases of Pulmonary Consumption," *Lancet* 12 (1829): 517–520, 517–8. On Smollett's Mediterranean travels and medicine, see E. Underwood, "Medicine and Science in the Writing of Tobias Smollett," *Proceedings of the Royal Society of Medicine* 30 (1937): 961–974; John F. Sena, "Smollet's Persona and the Melancholic Traveler," *Eighteenth-Century Studies* 1 (1968): 353–369.

68. Johnson, *Change of Air*, 258.

69. Johnson, *Change of Air*, 259.

70. K. Thompson, "Insalubrious California: Perception and Reality," *Annals of the Association of American Geographers* 59 (1969): 50–64.

71. See Paul P. Bernard, *Rush to the Alps: the Evolution of Vacationing in Switzerland* (New York: Columbia University Press, 1978); George Boddington, *An Essay on the Treatment and Cure of Pulmonary Consumption* (London: Longman, 1840); F.B. Rogers, "The Rise and Fall of the Altitude Therapy of Tuberculosis," *Bulletin of the History of Medicine* 43 (1969): 1–16; Hermann Weber, "On the influence of the Alpine climates on pulmonary consumption," *British Medical Journal* (1861): 41–42; 58–59; 148–149.

72. "Review: Dr Wilson's Statistical Reports on the Health of the Navy, for the Years 1830-1-2-3-4-5-6," *British and Foreign Medical Review* 11 (1840): 193.

73. Alexander Tulloch, *Statistical Reports of the Sickness, Mortality, and Invaliding among the Troops in the United Kingdom, Mediterranean,*

and the British America Presented to Both Houses of Parliament, by Command of her Majesty (London: W. Clowes, 1839); R.L. Blanco, "Henry Marshall (1775–1851) and the Health of the British Army," *Medical History* 4 (1970): 260–279; John Hennen, *Sketches of the Medical Topography of the Mediterranean* (London: Thomas Underwood, 1830). *Times* editors gave the issue wider attention on the paper's pages, with excerpts from the *Statistical Reports on the Sickness, Mortality, and Invaliding among the Troops on the Western Coast of Africa, the Island of St. Helena, the Cape of Good Hope, and the Mauritius, Times,* December 22, 1840, 7.

74. C.T. Williams, *The Climate of the South of France and It's Varieties Most Suitable for Invalids* (London: John Churchill and Sons, 1867); idem, *The Influence of Climate in the Prevention and Treatment of Pulmonary Consumption* (London: Smith, Elder, 1877).

75. J.A. Lindsay, *The Climatic Treatment of Consumption* (London: MacMillan, 1887), 2–3; for the opposite view of humidity, see Edwin Lee, *Nice and Its Climate* (London, W.J. Adams, 1855).

76. Lindsay, *The Climatic Treatment,* 40; Pemble, *Mediterranean Passion,* 93–5. See Walter Pagel, "Humoral Pathology: Alingering Anachronism in the History of Tuerculosis," *Medical History* 29 (1955): 299–308.

77. John C. Atkinson, *Change of Air: Fallacies Regarding It* (London: John Olivier, 1848); idem, *Practical Observations on Epidemic Cholera* (London: John Ollivier, 1848); idem, "On the Effects of Different Winds on the Human Constitution," *Lancet* 54 (1849): 207, 318; idem, "Electricity in Cholera," *Lancet* 54 (1849): 50; idem, "Fear—A Therapeutic Agent," *Lancet,* 58 (1851): 428; idem, *On Sleep, and Sleeplessness* (London: Trubner, 1867).

78. Lefevre, *Thermal Comfort,* passim.

79. Atkinson, *Change of Air,* 58.

80. Julius Jeffreys, *A Word on Climate and Atmospheric Influence* (London: Longman, 1850), and *A Few Remarks upon an Atmospheric Treatment of the Throat and Lungs* (London: Longman, 1847).

81. On astrometeorology, see Katharine Anderson, *Predicting the Weather* (Chicago: Chicago University Press, 2005).

82. Jeffreys, *A Word on Climate,* 6. For biography, see Andrew Marshall and Judith Marshall, *Striving for the Comfort Zone: A Perspective on Julius Jeffreys* (Dallas: Windy Knoll Publications, 2004).

83. Jeffreys, *A Word on Climate,* 13, 14. Respirators were meant to be used outdoors, but even homebound patients could put them on when their quarters were aired. The apparatus was described as enhancing the springiness of the lungs and the oxidation of the blood, and protected the user from the creation of tubercles. It was advertised to heal most bronchitis, whooping cough, and the effects of spasmodic and inflammatory croup and measles; ibid., 19, 20, 35.

84. *Times,* December 22, 1837, 8; Jeffrey advertised it in *Times,* August 11, 1838, 2, as "suitable for all ranks of society."

85. Garrett endorsed the use of the instrument in windy conditions because it stopped the direct impact of wind on the face, warmed the air, but most of all, because it condensed the exhaled humidity and so prevented the northeast wind from rubbing it off. C.B. Garret, *East and Northeast Winds* (London: Samuel Highely, 1855), 35, 67. Reviews were always complimentary. In one, the instrument was thought "productive of the greatest comfort to individuals with irritable air passages, enabling them to go into the open air in winter without suffering the pain of dyspnoea or cough to which they were otherwise subject in such circumstances." As demand increased, Jeffreys reduced the price to ten to twenty shillings, depending on the design and ornamentation. "Review [of C.B. Garret's *East and Northwest Wind*]," *British Medical and Foreign Report*, October 1840, 591.
86. Atkinson, *Change of Air*, 36–37.
87. Johnson, *Change of Air*, 257.
88. Atkinson, *Change of Air*, 43.
89. Atkinson, *Change of Air*, 23; see also James Riley, *The Eighteenth-Century Campaign to Avoid Disease* (New York: St, Martin's Press, 1987); see also Simon Schaffer, "Measuring Virtue: Eudiometry, Enlightenment, and Pneumatic Medicine," in A. Cunningham and R. French (eds.), *The Medical Enlightenment of the Eighteenth Century* (Cambridge: Cambridge University Press, 1990).
90. Atkinson, *Change of Air*, 27.
91. Atkinson, *Change of Air*, 28.
92. Thomas Burgess, *The Physiology or Mechanism of Blushing* (London; Churchill, 1839).
93. Thomas Burgess, "Inutility of Resorting to the Italian Climate for the Cure of Pulmonary Consumption," *Lancet* 1 (1850): 591–94; 2, 10–12; 525–27; 592; 700–703.
94. Thomas Burgess, *Climate of Italy in Relation to Pulmonary Consumption: With Remarks on the Influence of Foreign Climates upon Invalids* (London: Longman, 1852). Contemporary commentary on this phenomenon was profuse. "To seek in its sunny clime the renovation of exhausted nature—to repair ravages of insidious consumption, how many in beautiful Italy have thus found a grave! The cemeterie of this lovely land contain sad memorials of those who, in their attempt, have vainly sought its shores," "Review," *Bentley's Miscellany*, 32 (1852), 319.
95. Johnson, *Change of Air*.
96. *Phthisis*, 517.
97. Burgess, "Inutility," 592. See also Paul Cassar, *Medical History of Malta* (London: Wellcome, 1965).
98. John Adams, *A Short Account of the Climate of Madeira: With Instructions to Those Who Resort Thither for the Recovery of their Health* (London, T. Gillet, 1801); another complimentary account, speaking of lemons "of a monstrous size," is Thomas Ashe, *A Commercial*

View and Geographical Sketch, of the Brasils in South America, and of the Island of Madeira (London: Allen and Co., 1812).

99. William Gourlay, *Observations on the Natural History, Climate, and Disease of Madeira during the Period of 18 Years* (London: J. Callow, 1811), 31.

100. T. Wilson, (1842). "On the Principles and Treatment of Pulmonary Consumption," *Lancet* 37 (1842): 749–752, 752.

101. Kentish, *An Account of Baths and of a Madeira-House at Bristol*; U. Palmedo, "Exposition of a Method Employed for the Cure of Consumption." *Lancet* 37 (1841): 176–188, 176. On Heineken, see "Dr. Heineken's Meteorological Register Kept at Funchal, in Madeira, in the Year 1826; With Some Prefatory Observations on the Climate of That Island, &c," *Philosophical Magazine* 11 (1827): 362; 411. See also "Remarks on Madeira, Climate of the Tropics, Trade Winds, Rio Janeiro, the Polar Ice, etc." *Philosophical Magazine and Journal* 54 (1819): 107, 194. The idea was used as an advertisement for the application of steam-heating systems by Robertson Buchanan, *Practical and Descriptive Essays on the Economy of Fuel, and Management of Heat* (Glasgow: J. Hedderwick, 1810), 304.

102. Sir James Prior, *Narrative of a Voyage in the Indian Seas, in the Nisus Frigate, to the Cape of Good Hope, Isles of Bourbon, France, and Scychelles; to Madras; and the Isles of Java, St. Paul, and Amsterdam, during the Years 1810 and 1811* (London: R. Phillips, 1820), 2.

103. James Drummond Burns, "Porto Santo as Seen from the North of Madeira," in *The Vision of Prophecy* (London: James Nisbet, 1865).

104. The plan was to observe temperature, humidity, sunshine, pressure, wind, dew point, and precipitation. Climatotherapy as a science was to address "all these causes, which are constantly varying, carefully registered together with an account of topographical relations of the place of observation, the geological and mineralogical formations, constituting the basis of its soil, its state of cultivation, the most frequent winds to which it subject, the moral condition of its inhabitants, and the disease prevailing at the time—especially their various, types, we should make rapid progress in meteorological science, and possess positive data, with reference to the action of different atmospheric conditions on the animal economy both in healthy as well as in its diseased state," John Abraham Mason, *A Treatise on the Climate and Meteorology of Madeira*, James Sheridan Knowles (ed.) (London: Churchill, 1850), 177.

105. Mason, *A Treatise*, x–xi.

106. Mason, *A Treatise*, 133–4.

107. Pemble, *Mediterranean Passion*, 248.

108. Burgess, *Climate of Italy*, 20.

109. On spatial meaning of nostalgia, see Kevis Goodman, "'The Uncertain Disease': Nostalgia as a Sensuous Science of History," *Bloomington Workshop Reader*, 2007. Also see Lisa O'Sullivan,

"Place, Loss, and Longing: Clinical Nostalgia and the Boundaries of Identity in 19th-Century France" (London, PhD (Q.M. Hist.), 2007. Contemporary views on adaptation, providentiality, and climatological determinism are described in Robertson, *A General View of the Natural History of the Atmosphere*, passim; Harrison, *Climates and Constitutions*; Humboldt defined climate as a sum total of the forces whose effect on the constitution of organized beings "prohibit their permanent migration from one region of earth's surface to another," in R.E. Scoresby-Jackson, *Medical Climatology* (London: Churchill, 1862), 2.

110. Mason, *Treatise*, 115.
111. John Orton, "Effects of the Climate of Upper Canada on the Lungs." *Lancet* 40 (1840): 824–5, 825
112. White, *A Short Account of the Climate of Madeira*.
113. Burgess, *Climate of Italy*, 12
114. Burgess, *Climate of Italy*, 12, 18.
115. J. M. Bloxam, *The Climate of the Island of Madeira: and the Errors and Misrepresentations of Some Recent Authors on the Subject* (London, T. Richards, 1854), 1.
116. Bloxam, *The Climate*, 1, 92.
117. Bloxam, *The Climate*, 2.
118. Pemble, *Mediterranea Passion*, 246.
119. Bloxam, *The Climate*, 94.
120. T.M. Madden, *The Principal Health-Resorts of Europe and Africa for the Treatment of Chronic Diseases* (London: J. & A. Churchill, 1876), 48–49.
121. Lindsay, *Climatic Treatment*, 40–1; Scoresby-Jackson, *Medical Climatology*, 81.
122. Williams, *The Influence of Climate*, 81.
123. G.H. Brandt, "Notes on the Climate of Madeira," *Lancet* 87 (1866): 133.
124. Lindsay, *Climatic Treatment*, 195.
125. J. Andrews, "Letting Madness Range: Travel and Mental Disorder, c.1700–1900," in Wrigley and Revill (eds.), *Pathologies of Travel*, 36.
126. "Discomforts in Connexion with the Journey to and from Egypt," *Lancet* 154 (1899): 1776–77, 1776. An extensive treatment of the island from medical and tourist perspective was the four-part "The Island of Maderia," *The Lancet* 143 (1894): 55–59.
127. Christopher Hamlin, *Science of Impurity: Water Analysis in Nineteenth-Century Britain* (Berkeley: University of California Press, 1990), 332.
128. Pemble, *Mediterranean Passion*, 251.
129. On British spas and seaside resorts, see Roy Porter (ed.), *The Medical History of Waters and Spas* (Medical History, Supplement No. 10. London: Wellcome Institute for the History of Medicine, 1990);

J. V. N. Soane, *Fashionable Resort Regions: Their Evolution and Transformation* (Wallingford: Cab International, 1993).

130. Alfred Haviland, "The Manx Climate," *Lancet* 129 (1887): 1201–2.

131. "Review" in *Bentley's Miscellany* 32 (1852): 319.

132. William Harwood, *On the Curative Influence of the Southern Coast of England Especially That of Hastings with Observations on Diseases* (London: Colburn, 1828); James Mackness, *Hastings Considered as a Resort for Invalids* (London: Churchill, 1841); Thomas Shapter, *The Climate of the South of Devon* (London: John Churchill, 1842); C. R. Hall, *Torquay in Its Medical Aspect as a Resort for Pulmonary Invalids* (London: John Churchill, 1857); and many others.

133. Turner, *Taking the Cure*, 206.

134. F. Chambers, "Margate as a Health Resort," *Lancet* 84 (1864): 52.

135. Anthony D. King, *Colonian Urban Development. Culture, Social Power, and Environment* (London: Routledge and Kegan Paul, 1976).

136. August Bozzi Granville, *The Spas of England and Principal Sea-bathing Places* (London: H. Colburn, 1841), 526.

137. George Symes Hooper, *Observations on the Topography, Climate, and Prevalent Disease of the Island of Jersey* (London: Whittaker, 1837).

138. Michale Worboys, *Spreading Germs: Disease Theories and Medical Practice in Britain, 1865–1900* (Cambridge: Cambridge University Press, 2000), 228.

139. G.V. Poore "On Climate in Its Relation to Health," *The Lancet* 125 (1885): 212; Playfair, "Insanitary Conditions," 273.

140. William Gordon, W. *The Place of Climatology in Medicine* (London: H. K. Lewis, 1913), 19.

Conclusion

1. Douglas Gasking, "Causation and Recipes," *Mind* 64 (1955): 479–87; Calvin Woodard, "Reality and Social Reform: The Transition from Laissez-Faire to the Welfare State," *Yale Law Journal* 72 (1962): 286–328; and J. Willard Hurst, *Law and the Conditions of Freedom in the Nineteenth-Century United States* (Madison, 1956), 73.

2. Thornstein Veblen, *Conspicuous Consumption* (London: Penguin, [1899], 2005), 56.

3. Christopher Lasch, *The Culture of Narcissism: American Life in an Age of Diminishing Expectations* (London: Norton, 1991).

4. David Boswell Reid, *Illustrations of the Theory and Practice of Ventilation* (London: Longmans, 1844), 39.

Bibliography

Primary Sources

James Makittrick Adair, *Commentaries on the Principles and Practice of Physic* (London: T. Becket, 1772).

———, *Medical Cautions, for the Consideration of Invalids; those Especially who Resort to Bath: Containing Essays on Fashionable Diseases* (Bath: R. Cruttwell, 1786).

———, *Essay on Fashionable Diseases* (London: T. P. Bateman, 1790).

George Adams, *Lectures on Natural and Experimental Philosophy* (5 Volumes. London: R. Hidmarsh, 1794).

John Aitken, *Elements of the Theory and Practice of Physic and Surgery* (London, n.p., 1783).

James Anderson, *A Practical Treatise on Chimneys* (Edinburgh: C. Elliot, 1776).

Angeline: Or Sketches from Nature (3 vols. London: Kerby, 1794).

Annual Register for the Year 1765 (Fourth edition, London: J. Dodsley, 1784).

The Anti-Craftsman: Being an Answer to The Craftsman Extraordinary (London: Brindley, 1729).

John Arbuthnot, *An Essay Concerning the Effects of Air on Human Bodies* (London: J. Tonson, 1733).

Arley: Or the Faithless Wife (2 vols. London: J.S. Barr, 1790).

John Ash, *The New and Complete Dictionary of the English language* (London: Edward and Charles Dilly, 1775).

Augusta Denbeigh (Dublin: Brett Smith, 1795).

Girogio Baglivi, *The Practice of Physick, Reduc'd to the Ancient Way of Observations Containing a Just Parallel between the Wisdom and Experience of the Ancients* (London: Andrew Bell, 1704).

Edward Bancroft, *Remarks on the Review of the Controversy between Great Britain and Her Colonies. In Which the Errors of Its Author are Exposed* (London: T. Becket, 1769).

Robert Bath, *An Essay on the Medical Character* (Fifth Edition, London: C. Laidler, [1790?]).

Thomas Beddoes, *A Guide for Self-Preservation* (Bristol: Bulgin and Rosser, 1794).

———, *A Lecture Introductory to a Course of Popular Instruction on the Constitution and Management of the Human Body* (Bristol: N. Biggs, 1797).

———, *Manual of Health: or the Invalid Conducted Safely through the Seasons* (London: J. Johnson, 1806).

———, *Hygiea* (London: J. Mills, 1802).

———, *Essay on the Causes, Early Signs, and Prevention of Pulmonary Consumption for the Use of Parents and Preceptors* (London: Longman and Rees, 1799).

James Beresford, *The Miseries of Human Life or the Groans of Samuel Sensitive and Timothy Testy* (Sixth Edition, London: J. Ballantyne, 1807).

Walter Bernan, *On the History and Art of Warming and Ventilating Rooms and Buildings* (London: George Bell, 1845).

Adam Beuvius, *Henrietta of Grestenfeld* (London: William Lane, 1787–88).

Charles Bisset, *An Essay on Medical Constitution of Great Britain* (London: A. Millar, 1762).

Herman Boerhaave, *Dr. Boerhaave's Academical Lectures on the Theory of Physic* (6 Volumes. London: W. Innys, 1742–46).

Boerhaave's Aphorisms: Concerning the Knowledge and Cure of Diseases (Third Edition, London: W. Innys, 1755).

Richard Boulton, *Some Thoughts Concerning the Unusual qualities of the Air* (London: John Hooke, 1724).

Martha Bradley, *The British Housewife: or, the Cook, Housekeeper's, and Gardiner's Companion* (London, [1760?]).

Richard Brocklesby, *Reflections on Antient and Modern Musick, With the Application to the Cure of Diseases* (London: M. Cooper, 1749).

Robert Brookes, *The General Practice of Physic* (London: J. Newbery, 1754).

William Buchan, *Domestic Medicine: Or, a Treatise on the Prevention and Cure of Diseases by Regimen and Simple Medicines* (London: W. Strahan, 1772).

Thomas Burgess, "Inutility of resorting to the Italian Climate for the Cure of Pulmonary Consumption," *Lancet* 55 (1850): 591.

John Burton, *A Treatise on the Non-Naturals* (York: A. Staples, 1738).

William Cadogan, *A Dissertation on the Gout, and all Chronic Diseases* (London: J. Dodsley, 1771).

Louis La Caze, *Idee de l'homme physique et moral* (Paris: H. L. Guerin and L. F. Delatour, 1755).

Ephraim Chambers, *Cyclopaedia; or, An Universal Dictionary of Arts and Sciences*, Volume 2 (Dublin: J. Chambers, 1780).

Rice Charleton, *Cases of Patients Admitted into the Hospital at Bath, under the Care of the Late Dr. Oliver* (London: T. Cruttwell, 1776).

The Compleat Family Physician (Newcastle upon Tyne: printed by Matthew Brown, 1800–1801).

Bryan Cornwell, *The Domestic Physician; or, Guardian of Health* (London: J. Murray, 1784).

Jan Dalley, *The Black Hole: Money Myth, and Indian Empire* (London: Penguin, 2006).

Philip Dormer Stanhope Chesterfield, *Letters Written by the Late Right Honourable Philip Dormer Stanhope, Earl of Chesterfield, to His Son, Philip Stanhope* (4 Volumes. Fifth Edition, London: J. Dodsley, 1774).

George Cheyne, *A New Theory of Continual Fevers* (H. Newman, and J. Nutt, 1701).

———, *The English Malady* (London: G. Strahan, 1733).

———, *An Essay on Regimen* (London: C. Rivington and J. Leake, 1740).

———, *An Essay on Health and Long Life* (London: G. Strahan and J. Leake, 1724).

John Cheshire, *A Treatise Upon the Rheumatism* (London: C. Rivington, 1735).

Colin Chisholm, "On the Statistical Pathology of Bristol and Clifton," *Edinburgh Medical Journal* 13 (1817): 265–93.

Francis Clifton, *Hippocrates Upon Air, Water, and Situation* (London: J. Watts, 1734).

William Cobbett, *The Republican Judge: Or the American Liberty of the Press, as Exhibited, Explained, and Exposed* (Second Edition, London: J. Wright, 1798).

Jeremy Collier, "Of the Spleen," in *Essays Upon Several Moral Subjects* (London: J. and J. Knapton, 1731–32).

Anthony Collins, *A Philosophical Inquiry Concerning Human Liberty* (London: R. Robinson, 1717).

Emanuel Collins, *Lying Detected; or, Some of the Most Frightful Untruths That Ever Alarmed the British Metropolis, Fairly Exposed* (London: E. Farley, 1758).

Daniel Cox, *Observations on the Intermitting Pulse, as Prognosticating, in Acute Diseases* (London: A. Millar, 1758).

Country Gentleman, *A Narrative in Justification of Injured Innocence* (London: W. Webb, 1749).

James Graham, *A New and Curious Treatise of the Nature and Effects of Simple Earth, Water, and Air* etc. (London: Richardson, 1793).

Nicholas Culpeper, *The English Physician Enlarged* (London: A. and J. Churchill, 1708).

James Curry, *Popular Observations on Apparent Death from Drowning, Suffocation etc* (Northampton: T. Dicey, 1792).

Daily Advertiser, March 7 ("London Debates: 1789," in *London Debating Societies 1776–1799* [1994]).

Thomas Denman, *An Introduction to the Practice of Midwifery* (2 Volumes. London: J. Johnson, 1794–95).

A Discourse on the Doctrine of the Necessity of Human Actions (London: W. Bickerton, 1744).

Domestic Economy; or, Complete System of English Housekeeping (London: J. Creswick 1794).

Bartholomew di Dominiceti, *Medical Anecdotes of the Last Thirty Years* (London: L. Davis, 1781).

James Drake, *Anthropologia Nova* (London: Samuel Smith and Benjamin Walford, 1707).

Richard Drake, *An Essay on the Nature and Manner of Treating the Gout* (London: Author, 1758).

James Dunbar, *Essay on the History of Mankind in Rude and Cultivated Ages* (London: W. Strahan, 1781).

The Economist and General Adviser (London: Knigh and Lacey, 1825).

Encyclopædia Britannica (Edinburgh: A. Bell and C. Macfarquhar, 1771).

An Enquiry into the Causes of the Present Epidemical Diseases (London: F. Fayram, 1729).

Essay on the Most Rational Means of Preserving Health (London: James Wallis, 1799).

William Falconer, *Remarks on the Influence of Climate, Situation, Nature of Country, Population, Nature of Food, and Way of Life, on the Disposition and Temper* (London: C. Dilly, 1771).

William Falconer, *An Essay on the Preservation of the Health of Persons Employed in Agriculture* (London: R. Cruttwell, 1789).

William Falconer, *Observations Respecting the Pulse* (London: T. Cadell, 1796).

David Fordyce, *Dialogues Concerning Education* (2 vols. London: E. Dilly, 1757).

William Forster, *A Treatise on the Causes of Most Diseases Incident to Human Bodies* (London: J. Clarke, 1746).

John Fowler and John Cornforth, *English Decoration in the 18th Century* (London: Barrie and Jenkins, 1974).

Benjamin Franklin, *Observations on Smoky Chimneys* (London: John Debrett, 1787).

John Freind, *Nine Commentaries Upon Fevers* (London: T. Cox, 1730).

Foundling Hospital (London, England), *An Account of the Hospital for the Maintenance and Education of Exposed and Deserted Young Children* (London: n.p. 1759).

The Fountain of Knowledge, or, British Legacy (London: Bailey, 1760).

Hector Gavin, *Sanitary Ramblings* (London: John Churchill, 1848).

Leonard Gillispie, *Advice to the Commanders and Officers of His Majesty's Fleet Serving in the West Indies* (London: J. Cuthell, 1798).

Thomas Gordon, *Humorist: Being Essays Upon Several Subjects* (London: W. Boreham, 1720).

James Gregory, "Theory of Medicine, Notes of Lectures" (Edinburgh, ca. 1785), Wellcome Trust Library MS 2597.

James Gregory, *A Dissertation on the Influence of Change of Climate in Curing Diseases* (Philadelphia: T. Dobson, 1815).

John Gregory, *A Father's Legacy to His Daughters* (Edinburgh: A Strahan and T. Cadell, 1774).

Andrew Hamper, *The Economy of Health* (London: Author, 1785).

Jonas Hanway, *Serious Considerations on the Salutary Design of the Act of Parliament* (London: John Rivington, [1762]).

Jonas Hanway, *Midnight the Signal* (London: Dodsley, 1779).

William Heberden, "On the Influence of Cold upon the Health of the Inhabitants of London," *Philosophical Transactions* 86 (1796): 279–84.

John Hemet, *Contradictions, or Who Would Have Thought It* (London: Earl and Hemet, 1799).

John Hill, *Hypochondriasis. A Practical Treatise on the Nature and Cure of That Disorder; Commonly Called the Hyp and Hypo* (London: T. Trueman, 1775).

Hippocrates's Treatise on the Preservation of Health (London: John Bell, 1776).

Benjamin Hutchinson, *Biographia Medica* (London: J. Johnson, 1799).

John Huxham, *Observations of the Air and Epidemical Diseases, Made at Plymouth from 1728–1737* (Translated from Latin. London: J. Hinton, 1759).

The Idler by the Author of the Rambler (London: J. Rivington, 1790).

Robert Jackson, *An Outline of the History and Cure of Fever, Endemic and Contagious* (Edinburgh: Mundell and Son, 1798).

Seguin Henry Jackson, *A Treatise on Sympathy* (London: J. Murray, 1781).

James Johnson, *Change of Air or Pursuit of Health* (London: S. Highley, 1839).

Samuel Johnson, *A Dictionary of the English Language* (London: W. Strahan, 1755–56).

James Keill, *The Anatomy of the Humane Body Abridg'd* (London: Ralph Smith, 1703).

Nathaniel Lancaster, *Public Virtue: Or, the Love of Our Country* (London: R. Dodsley, 1746).

James Latta, *A Practical System of Surgery* (Edinburgh: G. Mudie, 1793).

John Leake, *A Course of Lectures on the Theory and Practice of Midwifery* (London: A. D. 1767).

John Leake, *Medical Instructions towards the Prevention and Cure of Chronic Diseases Peculiar to Women* (London: R. Baldwin, 1777).

John Locke, *An Essay Concerning Human Understanding*, Edited by P. D. Niditch (Oxford: Clarendon Press, 1975).

Bernard Lynch, *A Guide to Health through the Various Stages of Life* (London: Cooper, 1744).

J. Lyons, *Fancy-logy: A Discourse on the Doctrine of Necessity of Human actions, Proving It to Be a Fanaticism* (London: J. Purser, 1730).

Mariamne, or Irish Anecdotes (2 Volumes. Dublin: B. Smith, 1794).

James Mackenzie, *The History of Health* (Edinburgh: W. Gordon, 1758).

John Millar, *Observations on the Prevailing Diseases in Great Britain* (London: T. Cadell).

"Meteorological Observations [in Bristol, January 1774]," Wellcome Library MS. MSL. 111.

Giambattista Morgagni, *The Seats and Causes of Diseases Investigated by Anatomy* (3 Volumes. London: A. Millar, 1769).

George Motherby, *A New Medical Dictionary* (London: J. Johnson, 1775).

John Murray, MD, "Journal Containing Daily Meteorological and Monthly Medical Observations," Wellcome MS.7840.

A New System of Practical Domestic Economy: Founded on Modern Discoveries and the Private Communications of Persons of Experience (London: Colburn 1824).

Michael O'Ryan, *Advice, in the Consumption of the Lungs* (Dublin: H. Fitzpatrick, 1798).

Charles Perry, *Mechanical Account and Explication of the Hysteric Passion* (London: Shuckburgh, 1755).

Mr Phelps, *The Human Barometer: Or the Living Weather Glass. A Philosophic Poem* (London: M. Cooper, 1743).

John Prestwich, *Dissertation on Mineral, Animal, and Vegetable Poisons* (London: F. Newberry, 1775).

Richard Price, "Observation on the Difference between the Duration of Human Life in Towns and Country Parishes and Villages," *Philosophical Transactions* 65 (1683–1775): 424–445.

Joseph Priestley, *Experiments and Observations on Different Kinds of Air* (3 Volumes. London: J. Johnson, 1777).

Bernardino Ramazzini, *Health Preserved* (London: John Whiston, 1750).

Thomas Reid, *An Essay on the Nature and Cure of the Phthisis Pulmonalis* (London: T. Cadel, 1785).

Henry Robertson, *A General View of the Natural History of the Atmosphere* (Edinburgh: Abernethy and Walker, 1808).

Thomas Rowlandson, *The Comforts of a Modern Gala* (London: Thomas Tegg, 1807).

John Rutty, *A Chronological History of the Weather and Seasons and of the Prevailing Diseases in Dublin* (London: Robinson and Roberts, 1770).

Thomas Short, *A General Chronological History of the Air, Weather, Seasons, Meteors etc in Sundry Places and Different Times* (London: T. Longman and A. Millar, 1749).

Ebenezer Sibly, *The Medical Mirror* (London, [1800?]).

George Sigmond, "Materia Medica and Therapeutics," *Lancet* 30 (1837): 390–395.

John Sinclair, *The Code of Health and Longevity* (3 Volumes. Edinburgh: A. Constable, 1807).

Tobias George Smollett, *The Expedition of Humphry Clinker* (London: T. Johnston, 1771).

Tobias Smollett, *The Expedition of Humphry Clinker* (London: Harrison and Co., 1785).

Anton Freiherr von Störck, *A Second Essay on the Medicinal Virtues of Hemlock* (London: T. Becket, 1761).

[Samuel Strutt], *An Essay towards Demonstrating the Immateriality, and Free-agency of the Soul* (London: Shuckburgh, 1740).

Alexander Sutherland, *Attempts to Revive Antient Medical Doctrines. I. Of Waters in General. II. Of Bath and Bristol Waters in Particular. III. Of Sea Voyages* (London: A. Millar 1763).

James Stephen, *The Dangers of the Country* (London: J. Butterworth and J. Hatchard, 1807).

Percival Stockdale, *Three Discourses: Two Against Luxury and Dissipation. One on Universal Benevolence* (London: W. Flexney, 1773).

Thomas Sydenham, *The Entire Works of Dr Thomas Sydenham* (London: Edward Cave, 1742).

Robert Thornton, *The Philosophy of Medicine* (London: C. Whittingham, 1799).

———, *Medical Extracts* (3 Volumes. London: J. Johnson, 1796).

Charles Tomlison, *Rudimentary Treatise on Warming and Ventilation* (London: John Weale, 1850).

Jeremiah Wanewright, *Mechanical Account of the Non-Naturals* (London: Ralph Smith, 1707).

Joshua White, *Letters on England* (Philadelphia, for the Author, 1816).

John Whitehead, *The Life of the Rev. John Wesley, M.A. Some Time Fellow of Lincoln-College, Oxford* (London: Stephen Couchman, 1793–96).

Robert Whytt, *An Essay on the Vital and other Involuntary Motions of Animals* (Edinburgh: Hamilton, Balfour and Neill, 1751).

Robert Whytt, *Observations on the Nature, Causes, and Cure of those Disorders Which Have Been Commonly Called Nervous, Hypochondriac, or Hysteric* (Edinburgh: T. Beckett, 1765).

Thomas Willan, *Reports on the Diseases in London Particularly during the Years 1796, 1797, 1798, 1797, and 1800* (London: Phillips, 1801).

Anthony Florian Madinger Willich, *Lectures on Diet and Regimen: Being a Systematic Inquiry into the Most Rational Means of Preserving Health and Prolonging Life* (London: T. N. Longman, 1799).

Anthony Florian Madinger Willich, *The Domestic Encyclopaedia; or, A Dictionary of Facts, and Useful Knowledge* (4 Volumes. London: Murray and Highely, 1802).

Thomas Willis, *The Anatomy of the Brain* (London: T. Dring, 1681).

Clifton Wintringham, *Commentarium nosologicum morbos epidemicos et aeris variationes in urbe Eboracenci locisque vicinis, ab anno 1715, usque ad finem anni 1725* (Londini: J. Clarke, 1727).

John Wood, *A Series of Plans for Cottages or Habitations of the Labourer* (London: I. and J Taylor, 1792).

Dissertations

Robert Gusthart, *Specimen Medicum Inugurale de Aere ejusque in respiratione usu et effectibus* (Edinburgh: Thomas Ruddimann, 1740); Robert Willan *Dissertatio medica inauguralis de Qualitatibus Aeris*

(Edinburgh: James Cheyne, 1745); Ebenezer Macfait, *Dissertatio Medica Inauguralis de Aere, Aquis et Locis* (Edinburgh: 1745); John Gowdie, *Dissertatio medica inauguralis, de Aere quatenus morborum causa* (Edinburgh: Murray and Cohran, 1754); James Johnstone, *Tentamen Medicum Inaugurale de Aeris factitii imperio in primis corporis humani viis* (Edinburgh: T. and W. Ruddimannos, 1750); William Brown, *Specimen inaugurale pathologicum, de viribus atmosphaerae sentienti obviis* (Edinburgi: Apud Balfour, Auld, et Smellie, 1770), Daniel Rutherford, *Dissertation inauguralis de aere fixo dicto, aut mephitico, etc.* (Edinburgh: Balfour and Smellie, 1772); Edmundus Cullen, *Tentamen Medicum inugurale de Aere, et Imperio eju in Corpora Humana* (Edinburgh: Balfour and Smellie, 1781); Henri Burton, *Tentamen Physiologico-Medicum de Usu et Effectu Aeris Puri in Corpus Humanum.* (Edinburgh: Balfour and Smellie), 1788; Henry Robertson, *Dissertatio chemica medica inauguralis, de aere atmosphaerico* (Edinburgh: C. Stewart, 1801); William Cheekes, *Disputatio Chemica Inauguralis de Aere* (under George Baird) (Edinburgh: Adam Neill and Co., 1803); Nicholas Pitta, *Dissertatio physiologica inauguralis, de caeli effectu in genus humanum* (Edinburgh: C. Stewart, 1812); George Samuel Jenks, *Dissertatio medica inauguralis de coelo tabescentibus benigno* (Edinburgh: J. Pillans, 1821); William Jackson, *Tentamen chemica medica inauguralis de aëre communi* (Edinburgh: P. Neill, 1822).

Newspapers

British Foreign and Medical Review 15 (1843): 129.
Daily Gazetteer Saturday, January 12, 1740.
E. Johnson's British Gazette and Sunday Monitor (December 11, 1796).
General Evening Post, Thursday, January 17, 1740.
The London Magazine. Or, Gentleman's Monthly Intelligencer. London [England], [1747–83].
Morning Post and Daily Advertiser (October 25, 1776).
Oracle and Public Advertiser (March 13, 1795).
Stuart's Star and Evening Advertiser (March 24, 1789).
Sun (September 20, 1798).
Sun (December 28, 1799).
Telegraph (July 11, 1796).
The Times, May 10, 1802.
True Briton (March 27, 1793).
Johnson's British Gazette and Sunday Monitor (December 11, 1796).

Secondary Sources

D. G. C. Allan and R. E. Shoffield, *Stephen Hales: Scientist and Philanthropist* (London: Scholar Press, 1980).

Katharine Anderson, "Instincts and Instruments," in Christopher D. Green, Marlene Shore and Thomas Teo (eds.), *The Transformation of Psychology: Influences of 19th-Century Philosophy, Technology, and Natural Science* (Washington, DC: American Psychological Association, 2001).

H. Baer, M. Singer, and J. Johnsen, "Toward a Critical Medical Anthropology," *Social Science and Medicine* 34 (8): 95–98.

Luis Garcia-Ballester, *Galen and Galenism* (London: Ashgate/Variorum, 2002).

G. J. Barker-Benfield, *The Culture of Sensibility* (Chicago: Chicago University Press, 1996).

Maxine Berg, *Luxury and Pleasure in Eighteenth Century Britain* (Oxford: Oxford University Press, 2005).

Christopher J. Berry, " 'Climate' in the Eighteenth Century: James Dunbar and the Scottish Case," *Texas Studies in Literature and Language* 16 (1974): 281–92.

Jerome Blum, *The End of The Old Order in Rural Europe* (Princeton: Princeton University Press, 1978).

Barbara Brandon Schnorrenberg, "A True Relation of the Life and Career of James Graham, 1745–1794," *Eighteenth-Century Life* 15 (1991): 58–75.

Peter Brimblecombe, "Interest in Air Pollution among Early Fellows of the Royal Society," *Notes and Records of the Royal Society* 32 (1978): 123–29.

Laurence Brockliss and Colin Jones, *The Medical World of Early Modern France* (Oxford: Clarendon Press, 1997), 462.

Thomas Broman, "The Medical Science," in Roy Porter, *The Cambridge History of Science: Eighteenth-Century Science* (Cambridge: Cambridge University Press, 2003), 463–84.

Frank E. Brown, "Continuity and Change in the Urban House: Developments in Domestic Space Organisation in Seventeenth-Century London," *Comparative Studies in Society and History* 28 (1986): 558–90.

Theodore Browne, *The Mechanical Philosophy and the "Animal Economy"* (New York: Arno Press, 1981).

Bonnie Bullough, "The Causes of the Scottish Medical Renaissance of the Eighteenth Century," *Bulletin for the History of Medicine* 45 (1971): 13–28.

Andrew Burstein, "The Political Character of Sympathy," *Journal of the Early Republic* 21 (2001): 601–632.

Elizabeth Burton, *The Georgians at Home 1714–1830* (London: Longmans, 1967).

Jerome J. Bylebyl, "Galen on the Non-Natural Causes of Variations in the Pulse," *Bulletin of the History of Medicine* 45 (1971): 482–85.

W. F. Bynum, "Cullen and the Study of Fevers in Britain, 1760–1820," in W. F. Bynum and V. Nutton (eds.), *Theories of Fever from the Antiquity to the Enlightenment* (Medical Supplement No. 1, London: Wellcome Institute for the History of Medicine, 1981), 135–47.

Terry Castle, *The Female Thermometer: Eighteenth-Century Culture and the Invention of the Uncanny* (Oxford: Oxford University Press, 1995).

Carlo M. Cipolla, *Miasmas and Disease: Public Health and the Environment in the pre-Industrial Age* (Hew Haven and London: Yale University Press, 1992).

William Coleman, "Health and Hygiene in the Encyclopaedia: A Medical Doctrine for the Bourgeoisie," *Journal of the History of Medicine* (1974): 399–421.

William Coleman, *Death Is a Social Disease: Public Health and Political Economy in Early Industrial France* (Madison: Wisconsin University Press, 1982).

Alain Corbin, *The Foul and the Fragrant: Odor and the French Social Imagination* (Lemington Spa: Berg, 1986).

John Cornforth, *London Interiors: From the Archives of Country Life* (London: Aurum Press, 2000).

John Cornfort, *English Interiors, 1790–1848: the Quest for Comfort* (London: Barrie and Jenkins, 1978).

Daniel Cottom, "In the Bowels of the Novel: The Exchange of Fluids in the Beau Monde," *NOVEL: A Forum on Fiction* 32 (1999): 157–86.

John E. Crowley, *The Invention of Comfort: Sensibilities and Design in Early Modern Britain and Early America* (Baltimore and London: Johns Hopkins University Press, 2001).

Percy M. Dawson, "Stephen Hales, The Physiologist," *The Johns Hopkins Hospital Bulletin* 15 (1904): 1–15.

Michael Dorn, "Climate, Alcohol, and the American Body Politic: The Medical and Moral Geographies of Daniel Drake (1785–1852)," PhD Dissertation, University of Kentucky, 2003.

Oswald Doughty, "The English Malady of the Eighteenth Century," *The Review of English Studies* 2 (1926): 257–69.

Felix Driver, "Moral Geographies," *Transactions of the Institute of British Geographers* 25 (1988): 333–46.

Barbara Duden, *The Woman Beneath the Skin: A Doctor's Patients in Eighteenth Century Germany* (Cambridge, MA: Harvard University Press, 1998).

Markman Ellis, *The Politics of Sensibility: Race, Gender and Commerce in the Sentimental Novel* (Cambridge: Cambridge University Press, 1996).

Antoniette Emsch-Deriaz, "The Non-naturals Made Easy," in Roy Porter (ed.), *The Popularization of Medicine* (London and New York: Routledge, 1992), 134–59.

Antoinette S. Emsch-Deriaz, *Tissot: Physician of the Enlightenment* (New York: Peter Lang, 1992).

Oswald Doughty, "The English Malady of the Eighteenth Century," *The Review of English Studies* 2 (1926): 257–69.

Roger Ekirch, *At Day's Close: Night in Times Past* (New York: W. W. Norton, 2005).

Pedro Entralgo, *Mind and Body: Psychosomatic Pathology: A Short History of the Evolution of Medical Thought* (London: Harvill, 1955).

Robin Evans, *The Fabrication of Virtue: English Prison Architecture, 1750–1840* (Cambridge: Cambridge University Press, 1982).

David Ferriar, "The Erotics of Empiricism," unpublished presentation at University of Leeds, March 2, 2004.

Roger K. French, *Robert Whytt, The Soul and Medicine* (London: Wellcome Institute of the History of Medicine, 1969).

Northrop Frye, "Varieties of Eighteenth-century Sensibility," *Eighteenth-Century Studies* 24 (1990): 157–72.

Mark Girouard, *Life in the English Country House: A Social and Architectural History* (New Haven and London: Yale University Press, 1978).

Clarence J. Glacken, *Traces on the Rhodian Shore* (Stanford: University of California Press, 1990).

William Godwin, *Things as They Are* (London: G. G. and J. Robinson, 1796).

Jan Golinski, "The Human Barometer: Weather Instruments and the Body in Eighteenth-Century England," Paper given at the American Society for Eighteenth-Century Studies Annual Meeting, Notre Dame, Indiana, April 3, 1998.

———, "Barometers of Change: Meteorological Instruments as Machines of Enlightenment," in William Clark, Jan Golinski, and Simon Schaffer (eds.), *The Sciences in Enlightened Europe* (Chicago and London: Chicago University Press, 1999), 69–93.

———, *British Weather and the Climate of Enlightenment* (Chicago: University of Chicago Press, 2007).

Alan and Ann Gore, *The History of English Interiors* (Oxford: Phaidon, 1991).

Anita Guerrini, "Archibald Pitcairne and Newtonian Medicine," *Medical History* 31 (1987): 70–83.

David Hamilton, *The Healers: A History of Medicine in Scotland* (Edinburgh: Canongate, 1981).

Caroline Hannaway, "From Private Hygiene to Public Health: A Transformation in Western Medicine in the Eighteenth and Nineteenth centuries," in Teizo Ogawa (ed.), *Public Health: Proceedings of the 5th International Symposium on the Comparative History of Medicine* (Tokyo: Saikon, 1981), 108–28.

Geoffrey Harding, *Opiate Addiction, Morality and Medicine: From Moral Weakness to Pathological Disease* (Basingstoke: MacMillan, 1986).

J. Jean Hecht, *The Domestic Servant Class in Eighteenth-Century England* (London: Routledge & Kegan Paul, 1956).

Mieneke te Hennepe, *Depicting Skin: Visual Culture in Nineteenth-Century Medicine* (Wageningen: Ponsen and Looijen, 2007).

Bridget Hill, *Servants: English Domestics in the Eighteenth Century* (Oxford: Clarendon Press, 1996).

Dorothy Holland and Andrew Kipnis, "Metaphors for Embarrassment and Stories of Exposure: The not-so-Egocentric Self in American Culture," *Ethos* 22 (1994): 316–42.

Margaret Jacob, "The Mental Landscapes of the Public Sphere," *Eighteenth-Century Studies* (1994): 95–113.

Vladimir Janković, "The Nature of Place and the Place of Nature," *History of Science* 38 (2000): 79–113.

———, "Arcadian Instincts: a Geography of Truth in Georgian England," in Miles Ogborn and Charles W. J. Withers, *Georgian Geographies: Essays on Space, Place and Landscape in the Eighteenth Century* (Manchester: Manchester University Press, 2004), 174–91.

Saul Jarcho, "Galen's Six Non-naturals: A Bibliographic Note and Translation," *Bulletin for the History of Medicine* 44 (1970): 372–77.

Mark Jackson, *Health and the Modern Home* (London: Routledge, 2008).

Mark Jenner, "Underground, Overground: Pollution and Place in Urban History," *Journal of Urban History* 24 (1997): 97–110.

Lester King, "Some Problems of Causality in 18th Century Medicine," *Bulletin of the History of Medicine* 37 (1963): 15–24.

Christopher John Lawrence, "Medicine as Culture: Edinburgh and the Scottish Enlightenment," PhD Thesis, University College London, 1984.

Christopher Lawrence, "The Nervous System and Society in the Scottish Enlightenment," in Steven Shapin and Barry Barnes (eds.), *Natural Order* (London: Sage Publications, 1979), 19–40.

Ruth Leys, *From Sympathy to Reflex: Marshall Hall and His Opponents* (Garland: New York, 1990).

Dorothy Marshall, "Review G. E. and K. R. Fussel's *The English Countrywoman*," *Economic History Review* 7 (1954); 109–10.

Michael McKeon, "Aestheticising the Critique of Luxury," in Berg and Eger (eds.), *Luxury in the Eighteenth Century*, 57–70.

Tim Meldrum, "Domestic Service, Privacy and the Eighteenth-century Metropolitan Household," *Urban History* 26 (1999): 27–39.

Gregg Mitman, Michelle Murphy and Christopher Sellers, "Introduction: A Cloud Over History," *Osiris* 19 (2004): 1–17.

Sergio Moravia, "From Homme Machine to Homme Sensible: Changing Eighteenth-Century Models of Man's Image," *Journal of the History of Ideas* 39 (1978): 45–60.

Edmund Newell, "Atmospheric Pollution and the British Industry, 1690–1920," *Technology and Culture* 38 (1997): 655–89.

Peter H. Niebyl, "The Non-naturals," *Bulletin for the History of Medicine* 45 (1971): 486–92.

Felicity Nussbaum and Laura Brown (eds.), *The New Eighteenth Century* (Methuen: London 1987).

Gail Kern Paster, *The Body Embarrassed: Drama and the Disciplines of Shame in Early Modern England* (Ithaca, NY: Cornell University Press, 1993).

Severine Pilloud and Micheline Louis-Courvoisier, "The Intimate Experience of the Body in the Eighteenth Century: between Interiority and Exteriority," *Medical History* 47 (2003): 451–72.

George Pitcher, "Necessitarianism," *The Philosophical Quarterly* 11 (July 1961): 201–12.

Gianna Pomata, *Contracting a Cure: Patients, Healers and the Law in Early Modern Bologna* (Baltimore: Johns Hopkins University Press, 1998.

Dorothy Porter, "The Healthy Body in the Twentieth Century," in Pickstone and Cooter, *Medicine in the 20th century* (London: Routledge 2002), 201–206.

Roy Porter, "Lay Medical Knowledge in the Eighteenth Century: the Evidence of the *Gentleman's Magazine*," *Medical History* 29, No. 2 (1985).

Roy Porter, "Diseases of Civilization," in W. F. Bynum and Roy Porter (eds.), *Companion Encyclopedia of the History of Medicine* (London: Routledge, 1993), 585–600.

Roy Porter, "Nervousness, Eighteenth and Nineteenth Century Style: From Luxury to Labour," in Marijke Gijswijt-Hofstra and Roy Porter (eds.), *Cultures of Neurasthenia from Beard to the First World War* (Amsterdam: Rodopi, 2001), 31–47.

Roy Porter, *Quacks: Fakers and Charlatans in Medicine* (London: Tempus, 1989).

F. N. L. Poynter, "Sydenham's Influence Abroad," *Medical History* 17 (1973): 223–34.

L. J. Rather, "The Six Things Non-Natural: A Note on the Origins and Fate of a Doctrine and a Phrase," *Clio Medica* 3 (1968): 337–47.

Roselyne Rey, "Vitalism, Disease and Society" *Clio Medica* 29 (1995): 274–88.

George Rosen, "The Fate of the Concept of Medical Police," *Centaurus* 5 (1957): 97–113.

Christine Mesiner Rosen and Joel Arthur Tarr, "The Importance of an Urban Perspective in Environmental History," *Journal of Urban History* 20 (1994): 299–310.

Charles Rosenberg (ed.), *Right Living: An Anglo American Tradition of Self-Help Medicine and Hygiene* (Baltimore: John Hopkins University Press, 2003).

Lisa Rosner, *Medical Education in the Age of Improvement: Edinburgh Students and Apprentices, 1760–1826* (Edinburgh: Edinburgh University Press, 1991).

George Rousseau, "Nerves Spirits and Fibres: Towards Defining the Origins of Sensibility." *The Blue Guitar* 2 (1976): 125–53.

George Rousseau "Nerves Spiritis and Fibres: Toward Defining the Origins of Sensibility," in R. F. Brissenden and J. C. Eade (eds.), *Studies in the Eighteenth century III: Papers Presented at the Third David Nichol Smith Memorial Seminar, Canberra, 1973* (Toronto: University of Toronto Press, 1976).

Charles J. Rzepka, "De Quincey and Kant," *PMLA* 115 (2000): 93–94.

Sharon Ruston, *Shelley and Vitality* (London: Palgrave Macmillan, 2005).

Rafaella Sarti, *Europe at Home: Family and Material Culture 1500–1800* (New Haven: Yale University Press, 2002).

John Sekora, *Luxury: The Concept in Western Thought, Eden to Smollett* (Baltimore and London: Johns Hopkins University Press, 1977).

William Seller, "Memoirs of the Life and Writings of Robert Whytt," *Transactions of the Royal Society of Edinburgh* 23 (1864): 99–131.

Simon Schaffer, "Enlightened Automata," in William Clark, Jan Golinski, and Simon Schaffer (eds.). *The Sciences in Enlightened Europe* (Chicago: Chicago University Press, 1999), 126–65.

Carole Shammas, "The Domestic Environment in Early Modern England and America," *Journal of Social History* 14 (1980): 4–24.

Richard Brinsley Sheridan, *The School for Scandal* (London: E. Powell, 1798).

David E. Shuttleton, "'A Modest Examination': John Arbuthnot and the Scottish Newtonians," *British Journal for the Eighteenth-Century Studies* 18 (1995): 47–62.

Richard Shweder and Edmund Bourne, "Does the Concept of Person Vary Cross Culturally," in R. Shweder and R. LeVine (eds.), *Culture Theory* (Cambridge: Cambridge University Press, 1984).

Ginnie Smith, "Prescribing the Rules of Health: Self-Help and Advice in the Late Eighteenth Century," in Roy Porter (ed.), *Patients and Practitioners* (Cambridge: Cambridge University Press, 1985), 249–82.

Virginia Sarah Smith, *Clean: A History of Personal Hygiene and Purity* (Oxford: Oxford University Press, 2007).

Woodruff D. Smith, "Complications of the Commonplace: Tea, Sugar, and Imperialism," *Journal of Interdisciplinary History* 23 (1992): 259–78.

Barbara Maria Stafford, *Body criticism: Imaging the Unseen in Enlightenment Art and Medicine* (Cambridge, MA: MIT Press, 1991).

Lawrence Stone, "The Private and the Public in the Stately Homes in England, 1500–1990," *Social Research* 58 (1991): 227–57.

James Stephen Taylor, "Philanthropy and Empire: Jonas Hanway and the Infant Poor of London," *Eighteenth-Century Studies* 12, No. 3 (1979): 285–305.

Mary Terrall, "Metaphysics, Mathematics and the Gendering of Science in Eighteenth century France," in Clark, Golinski and Schaffer (eds.), *The Sciences in Enlightened Europe* (Chicago: University of Chicago Press, 1999), 246–71.

Anne Vila, *Enlightenment and Pathology: Sensibility in the Literature and Medicine of Eighteenth-Century France* (London: Johns Hopkins University Press, 1998).

Hans-Joachim Voth, "Time and Work in Eighteenth-Century London," *The Journal of Economic History* 58 (1998): 29–58.

Sara Warneke, "A Taste for Newfangledness: The Destructive Potential of Novelty in Early Modern England." *Sixteenth Century Journal* 26 (1995): 881–95.

Andrew Wear, "Making Sense of Health and the Environment in Early Modern England," in Andrew Wear (ed.), *Medicine in Society: Historical Essays* (Cambridge: Cambridge University Press, 1992), 119–47.

B. White, "Scottish Medicine and the English Public Health," in Derek Dow (ed.), *The Influence of Scottish Medicine* (Carnfort: Pantheon, 1988).

Elizabeth A. Williams, *The Physical and the Moral: Anthropology, Physiology, and Philosophical Medicine in France, 1750–1850* (Cambridge: Cambridge University Press, 1994).

Elizabeth A. Williams, "Hysteria and the Court Physician in Enlightenment France," *Eighteenth-Century Studies* 35 (2002): 247–55.

Raymond Williams, *The Country and the City* (London: Chatto and Windus, 1973).

Charles T. Wolfe and Motoichi Terada, "The Animal Economy as Object and Program in Montpellier Vitalism," *Science in Context* 21, No. 4 (2008): 537–79.

Gideon Yaffe, *Manifest Activity: Thomas Reid's Theory of Action* (Oxford: Oxford University Press, 2004).

Paul Youngquist, "De Quince's Crazy Body," *PMLA* 114 (1999): 346–58.

Index

Bradley, Marta, 59
Brockliss, Laurence, 26
Brown, Thomas, 24
Brydone, Patrick, 132
Buchan, William, 69, 74, 76,
   100, 108
Burgess, Thomas, 140 *passim*

Cadogan, William, 103
candles, 49, 51
Canguilhem, Georges, 10
carbon dioxide, 139
Carter, John, 72
causes of disease
   exciting, 122, 123
   external, 95
   immediate, 95
   non-necessary, 95
   predisposing, 30, 87, 123
   procatarctic, 95
   remote, 95
Chabannes, Marquis J. B. M. F.,
   83, 89
Chambers, William, 74
change of scene, 147
Cheyne, George, 20–1
chilblains, 59
chills, 30, 36, 60, 76, 96
Clark, James, 127–8, 136, 142
cleanliness, 62, 98, 152
cleanliness of dress, 104
climate
   and animal economy, 143–4
   artificial, 81, 124, 138
   chilly, 141
   choice of, 129 *passim*
   dangerous, 120
   and fibers, 24
   fomenting, 138
   genial, 124
   healthy, 127
   hereditary, 144
   Italian, 140
   maritime, 74
   miseries of, 124
   national, 74

native, 48
portable, 137, 138
suitable for Britons, 149
sultry, 134
temperate, 75, 134
uniform, 105
variable, 60
climatotherapy, as science, 145
climatotherapy, rationales of, 126
clothed body, 97
clothes, warm, 97
clothing, 94
Cobbett, William, 99
Cockburn, William, 29
cold bathing, 103
comfort, 4–5, 139
commercialization, 22, 43
common cold, 95
constipation, 59
consumption, disease, 16, 25, 28,
   47, 99, 110, 120, 125, 127,
   130–50
Corbin, Alain, 2
cotton, 98
country air, 7, 45, 47, 54
Coleman, William, 65
Cooke, William, 70
Crabbe, George, 35
crowding, 46, 51, 90
Crowley, John, 23, 56
Cullen, William, 47, 68
culture of narcissism, 154

dampness, 53, 73, 76, 99, 136, 145,
   see also air, moist
Davy, Humphrey, 130
delicacy, 27–8, 31, 124, 151
Desaguliers, John Theophilus, 69
Dillon, Michael, 4
discomfort, 46, 55–7, 89, 151
Douglas, Mary, 6
drafts, 61, 77, 106
dress
   and weather changes, 110
   female, 110
Duden, Barbara, 55